The Logica Yearbook

2022

The Logica Yearbook
2022

Edited by

Igor Sedlár

ISBN 978-1-84890-446-0

College Publications
Scientific Director: Dov Gabbay
Managing Director: Jane Spurr

www.collegepublications.co.uk

Original cover design by Laraine Welch

Preface

This volume contains peer-reviewed papers based on selected contributions presented at the conference *Logica 2022*, which took place in the Teplá Monastery, Czech Republic, on 11–15 September 2022.

Since 1987 the Logica conference series consists of meetings typically taking place in a quiet venue housing all participants together, and fostering fruitful exchanges between logicians of different specializations and generations, including students.

The programme of Logica 2022 comprised more than thirty lectures, including those given by our distinguished invited speakers Steve Awodey, Alena Vencovská, and Yde Venema. A Special Session on Logic and Information was part of the programme, consisting of invited talks given by Marta Bílková, Alessandra Palmigiano, and Heinrich Wansing.

On behalf of the Logica co-chair Vít Punčochář and the whole Logica organising team, I would like to thank the Institute of Philosophy of the Czech Academy of Sciences for its support of the Logica conference series, the staff of Teplá Monastery Hotel for their hospitality and friendly assistance, and the Czech Science Foundation for financial support (grant no. 21-23610M). Vít and I are grateful to the members of the programme committee for evaluating the abstracts submitted to the conference, to Kateřina Krusová for administrative and practical assistance, and to Ondrej Majer for designing the Logica T-shirts and other materials. As the editor of this volume, I am grateful to the reviewers of the papers for their time and valuable advice, and to College Publications and its managing director, Jane Spurr, for our pleasant cooperation during the preparation of this volume. Last but not least, I would like to thank authors of the papers included in this volume for their contributions and collaboration during the editorial process.

Prague, December 2023   Igor Sedlár

Table of Contents

On Duality Between Partiality and Paraconsistency

VINCENT DEGAUQUIER

Abstract: This article is intended to provide a unified understanding of some partial and paraconsistent logics through an investigation of their duality relations. In this connection, three many-valued logics and three constructive logics are addressed. Among the many-valued logics, we consider Kleene's strong three-valued logic, Priest's logic of paradox, and Dunn-Belnap's four-valued logic. Among the constructive logics, we investigate intuitionistic logic, dual-intuitionistic logic, and bi-intuitionistic logic. A general principle of duality is identified and shown to apply to each of these logics.

Keywords: temporal logic, many-valued logic, constructive logic, partial logic, paraconsistent logic, relational semantics, sequent calculus, duality

1   Introduction

Partiality and paraconsistency are metatheoretical properties attributed to certain non-classical logics. A logic is described as partial if it does not obey the law of excluded middle and it is described as paraconsistent if it does not obey the law of non-contradiction. These laws are inherited from a long philosophical tradition and give rise to various interpretations expressible in the language of formal logic (see Restall, 2004). In this discussion, they have the following meaning. The law of excluded middle states that, for every finite sequence of formulae Γ and for every formula A, the sequent $\Gamma \vdash (A \text{ or not } A)$ is valid. The law of non-contradiction states that, for every finite sequence of formulae Δ and for every formula A, the sequent $(A \text{ and not } A) \vdash \Delta$ is valid. Consequently, a logic is called partial if, for some finite sequence of formulae Γ and for some formula A, the sequent $\Gamma \vdash (A \text{ or not } A)$ is not valid and a logic is called paraconsistent if, for some finite sequence of formulae Δ and for some formula A, the sequent $(A \text{ and not } A) \vdash \Delta$ is not valid.

It should be noted that we depart from the standard definition of paraconsistency. Usually a logic is called paraconsistent if it does not obey the

1

Vincent Degauquier

principle of explosion, which states that, for all formulae A and B, the sequent $(A \text{ and not } A) \vdash B$ is valid. Insofar as this principle is a special case of the law of non-contradiction (i.e. the one where the sequence Δ consists of one and only one formula), we see that any logic which is paraconsistent in the usual sense is also paraconsistent in the sense we have just defined it.

Although the notions of partiality and paraconsistency can be made precise by defining them in terms of classes of valid sequents, their deep meaning remain strangely obscure. Are there model-theoretic or proof-theoretic features explaining why some sequent of the form $\Gamma \vdash (A \text{ or not } A)$ or $(A \text{ and not } A) \vdash \Delta$ are not valid according to a given logic? In other words, are there underlying model-theoretic or proof-theoretic features that would be specific to partial or paraconsistent logics?

This issue sounds even more relevant when we notice that the generic name 'partial logic', just like 'paraconsistent logic', covers a range of logics that are very different in nature. For example, both intuitionistic logic and Kleene's strong three-valued logic are partial in the sense mentioned above. Yet we know from an argument of Gödel (1986) that intuitionistic logic cannot be understood as a finitely-valued logic. Indeed, the model-theoretic interpretations of intuitionistic logic make use of a notion of model that is usually either topological or relational and that cannot be expressed by means of a function from the set of formulae to a finite set of truth-values.

The present discussion is intended to provide a unified understanding of some partial and paraconsistent logics through an investigation of their duality relations. In this connection, three many-valued logics and three constructive logics are addressed. Among the many-valued logics, we consider Kleene's strong three-valued logic, Priest's logic of paradox, and Dunn-Belnap's four-valued logic, also known as first-degree entailment logic. Among the constructive logics, we investigate intuitionistic logic, dual-intuitionistic logic, and bi-intuitionistic logic, also known as Heyting-Brouwer logic.

To propose a unified understanding of these partial and paraconsistent logics as well as a study of their duality relations, three steps mark out this article. First, we define a four-valued extension of the temporal logic $\mathsf{K_t T4}$. This logic provides a general framework from which the aforementioned logics can be investigated. According to the view that model-theoretic and proof-theoretic approaches are complementary and necessary for the complete characterisation of a logic, a relational semantics and a labelled sequent calculus are set out. Second, we identify three principles of duality using this four-valued extension of $\mathsf{K_t T4}$ and we show several properties associated with each of these principles. Among other things, we show that the combination of two of these principles gives rise to a well-known principle of duality between partiality and paraconsistency. Third, we point out that bi-intuitionistic logic and Dunn-Belnap's four-valued logic can be faithfully embedded into the four-valued extension of $K_t T4$. A general principle of duality is then identified and shown to apply to every partial or paraconsistent logic discussed.

2   A four-valued extension of $K_t T4$

The temporal logic $K_t T4$ is the modal logic obtained from the minimal temporal logic K_t (see Rescher & Urquhart, 1971) by requiring the accessibility relation to be both reflexive (which corresponds to the axiom T) and transitive (which corresponds to the axiom 4). This section aims to provide a relational semantics and a labelled sequent calculus for an extension of $K_t T4$ based on Dunn-Belnap's four-valued logic (see Belnap, 1977). This many-valued modal logic is referred to as $K_t^4 T4$ (see Degauquier, 2018).

2.1   Relational semantics

The *language* of $K_t^4 T4$, denoted by $\mathcal{L}(K_t)$, is composed of a countable set of propositional symbols p_n for every $n \in \mathbb{N}$ plus the propositional logical symbols $\neg$, $\wedge$, $\vee$ and the modal logical symbols $\Box_F$, $\Diamond_F$, $\Box_P$, $\Diamond_P$ (where 'F' stands for 'future' and 'P' stands for 'past'). The formulae of $\mathcal{L}(K_t)$ are defined as follows:

$$A ::= p \mid \neg A \mid (A \wedge A) \mid (A \vee A) \mid \Box_F A \mid \Diamond_F A \mid \Box_P A \mid \Diamond_P A$$

A *frame* $\mathcal{F}$ is a structure $\langle W, R \rangle$ in which W is a non-empty set (of possible worlds) and R is an ordered pair $\langle R_F, R_P \rangle$ such that R_F is a reflexive and transitive binary relation on W and R_P is the inverse relation of R_F. Note that it follows immediately from this definition that R_P is also reflexive and transitive.

A *model* $\mathcal{M}$ for $\mathcal{L}(K_t)$ is a structure $\langle W, R, V \rangle$ such that $\langle W, R \rangle$ is a frame and V is an ordered pair $\langle V^+, V^- \rangle$ such that V^+ and V^- are mappings from natural numbers to subsets of W. Thereby $V^+(n)$ denotes the set of possible worlds that verify the proposition p_n and $V^-(n)$ denotes the set of possible worlds that falsify the proposition p_n, for every $n \in \mathbb{N}$.

The *truth* and the *falsehood* of a formula of $\mathcal{L}(K_t)$ are defined at a world in a model. Given a world α in a model $\mathcal{M} = \langle W, R, V \rangle$, the truth (denoted

Vincent Degauquier

by $\mathcal{M}, \alpha \vDash^+$) and the falsehood (denoted by $\mathcal{M}, \alpha \vDash^-$) of the formulae of $\mathcal{L}(\mathsf{K_t})$ at α in $\mathcal{M}$ are defined inductively:

$\mathcal{M}, \alpha \vDash^+ p_n$ iff $\alpha \in V^+(n)$, for $n \in \mathbb{N}$
$\mathcal{M}, \alpha \vDash^- p_n$ iff $\alpha \in V^-(n)$, for $n \in \mathbb{N}$
$\mathcal{M}, \alpha \vDash^+ \neg A$ iff $\mathcal{M}, \alpha \vDash^- A$
$\mathcal{M}, \alpha \vDash^- \neg A$ iff $\mathcal{M}, \alpha \vDash^+ A$
$\mathcal{M}, \alpha \vDash^+ (A \wedge B)$ iff $\mathcal{M}, \alpha \vDash^+ A$ and $\mathcal{M}, \alpha \vDash^+ B$
$\mathcal{M}, \alpha \vDash^- (A \wedge B)$ iff $\mathcal{M}, \alpha \vDash^- A$ or $\mathcal{M}, \alpha \vDash^- B$
$\mathcal{M}, \alpha \vDash^+ (A \vee B)$ iff $\mathcal{M}, \alpha \vDash^+ A$ or $\mathcal{M}, \alpha \vDash^+ B$
$\mathcal{M}, \alpha \vDash^- (A \vee B)$ iff $\mathcal{M}, \alpha \vDash^- A$ and $\mathcal{M}, \alpha \vDash^- B$
$\mathcal{M}, \alpha \vDash^+ \Box_F A$ iff for all w in W, $\langle \alpha, w \rangle \in R_F$ implies $\mathcal{M}, w \vDash^+ A$
$\mathcal{M}, \alpha \vDash^- \Box_F A$ iff for some w in W, $\langle \alpha, w \rangle \in R_F$ and $\mathcal{M}, w \vDash^- A$
$\mathcal{M}, \alpha \vDash^+ \Diamond_F A$ iff for some w in W, $\langle \alpha, w \rangle \in R_F$ and $\mathcal{M}, w \vDash^+ A$
$\mathcal{M}, \alpha \vDash^- \Diamond_F A$ iff for all w in W, $\langle \alpha, w \rangle \in R_F$ implies $\mathcal{M}, w \vDash^- A$
$\mathcal{M}, \alpha \vDash^+ \Box_P A$ iff for all w in W, $\langle \alpha, w \rangle \in R_P$ implies $\mathcal{M}, w \vDash^+ A$
$\mathcal{M}, \alpha \vDash^- \Box_P A$ iff for some w in W, $\langle \alpha, w \rangle \in R_P$ and $\mathcal{M}, w \vDash^- A$
$\mathcal{M}, \alpha \vDash^+ \Diamond_P A$ iff for some w in W, $\langle \alpha, w \rangle \in R_P$ and $\mathcal{M}, w \vDash^+ A$
$\mathcal{M}, \alpha \vDash^- \Diamond_P A$ iff for all w in W, $\langle \alpha, w \rangle \in R_P$ implies $\mathcal{M}, w \vDash^- A$

Several semantic approaches can be specified according to the class of models considered. Two of these approaches seem particularly relevant for our purposes, namely the gappy semantics and the glutty semantics. These two types of semantics can be distinguished by defining some conditions on the models.

Let $\mathcal{M}$ be a model such that $\mathcal{M} = \langle W, R, V \rangle$. Then, $\mathcal{M}$ is *consistent* if $V^+(n) \cap V^-(n) = \emptyset$, for every $n \in \mathbb{N}$ and $\mathcal{M}$ is *complete* if $V^+(n) \cup V^-(n) = W$, for every $n \in \mathbb{N}$. In this sense, the model $\mathcal{M}$ is called *classical* if it is both consistent and complete.

Depending on whether a semantics restricts the class of models to that of consistent or complete models, this semantics will be called *gappy* or *glutty*, respectively. The reason why we call these semantics gappy or glutty lies in the fact that they do not obey the 'metalinguistic' law of excluded middle (stating that any sentence of the object-language has at least one of the values true and false) or the 'metalinguistic' law of non-contradiction (stating that any sentence of the object-language has at most one of the values true and false), respectively (see Dunn, 2000). By induction on the complexity of formulae, we obtain:

4

Proposition 1 (meta-law of excluded middle)   *Let $\mathcal{M}$ be a complete model for $\mathcal{L}(\mathsf{K_t})$. Then, for all formulae A of $\mathcal{L}(\mathsf{K_t})$ and for all worlds w in $\mathcal{M}$, $\mathcal{M}, w \vDash^+ A$ or $\mathcal{M}, w \vDash^- A$.*

Proposition 2 (meta-law of non-contradiction)   *Let $\mathcal{M}$ be a consistent model for $\mathcal{L}(\mathsf{K_t})$. Then, for all formulae A of $\mathcal{L}(\mathsf{K_t})$ and for all worlds w in $\mathcal{M}$, $\mathcal{M}, w \nvDash^+ A$ or $\mathcal{M}, w \nvDash^- A$.*

2.2   Labelled sequent calculi

The labelled sequent calculi described hereafter are based on an internalisation of the relational semantics of $\mathsf{K_t}\mathsf{T}4$ into a four-sided sequent calculus closely related to those developed by Girard (1976), Bochman (1998), and Muskens (1999). Similar approaches in the context of two-sided sequent calculi have been discussed, among others, by Bonnette and Goré (1998) as well as Negri (2005).

A *labelled sequent* Λ is a finite set of labelled formulae and structural elements. A *labelled formula* is a triple $\langle A, \lambda, x \rangle$ such that A is a formula of $\mathcal{L}(\mathsf{K_t})$, $\lambda \in \{0^-, 1^+, 0^+, 1^-\}$ and x is a natural number. A *structural element* is an ordered pair $\langle x, y \rangle$ where x and y are natural numbers. If Λ_1 and Λ_2 are labelled sequents and l is a labelled formula or a structural element, the labelled sequents $\Lambda_1 \cup \Lambda_2$ and $\{l\}$ are respectively denoted by Λ_1, Λ_2 and l.

Intuitively, a labelled formula $\langle A, \lambda, x \rangle$ means: 'Formula A has the minimum degree of falsehood (or, equivalently, is not false) at possible world x' if $\lambda = 0^-$; 'Formula A has the maximum degree of truth (or, equivalently, is true) at possible world x' if $\lambda = 1^+$; 'Formula A has the minimum degree of truth (or, equivalently, is not true) at possible world x' if $\lambda = 0^+$; 'Formula A has the maximum degree of falsehood (or, equivalently, is false) at possible world x' if $\lambda = 1^-$. In the same way, a structural element $\langle x, y \rangle$ means: 'Possible world y is accessible from possible world x'.

A labelled sequent Λ is *valid* if there is no counter-model to Λ. A model $\mathcal{M} = \langle W, R, V \rangle$ is a *counter-model* to Λ if there is a function $f : \mathbb{N} \to W$ such that:

- for every labelled formula $\langle A, \lambda, x \rangle$ in Λ:

    - $\mathcal{M}, f(x) \nvDash^- A$ if $\lambda = 0^-$
    - $\mathcal{M}, f(x) \vDash^+ A$ if $\lambda = 1^+$
    - $\mathcal{M}, f(x) \nvDash^+ A$ if $\lambda = 0^+$

5

Vincent Degauquier

- $\mathcal{M}, f(x) \vDash^{-} A$ if $\lambda = 1^{-}$

• for every structural element $\langle x, y \rangle$ in Λ, $\langle f(x), f(y) \rangle \in R_{\mathrm{F}}$.

This definition of validity can be preserved for the gappy and the glutty semantics. Depending on whether the notion of valid labelled sequent is restricted to consistent models or to complete models, a labelled sequent is called *gap-valid* or *glut-valid*, respectively. If only the class of classical models is taken into account, then a labelled sequent is called *classic-valid*.

To define labelled sequent calculi which are sound and complete with respect to these model-theoretic notions, rules of inference governing the behavior of the logical connectives are to be set out. It is worth noting that these calculi are free of weakening and contraction structural rules. We could have also defined labelled sequents as multisets and shown that these rules are admissible. Instead, we preferred to start with labelled sequents as sets and avoid this exercise.

$$\frac{\Lambda, \langle A, 1^{-}, x \rangle}{\Lambda, \langle \neg A, 1^{+}, x \rangle} \ 1^{+} \neg \qquad\qquad \frac{\Lambda, \langle A, 0^{-}, x \rangle}{\Lambda, \langle \neg A, 0^{+}, x \rangle} \ 0^{+} \neg$$

$$\frac{\Lambda, \langle A, 0^{+}, x \rangle}{\Lambda, \langle \neg A, 0^{-}, x \rangle} \ 0^{-} \neg \qquad\qquad \frac{\Lambda, \langle A, 1^{+}, x \rangle}{\Lambda, \langle \neg A, 1^{-}, x \rangle} \ 1^{-} \neg$$

$$\frac{\Lambda, \langle A, 1^{+}, x \rangle, \langle B, 1^{+}, x \rangle}{\Lambda, \langle (A \wedge B), 1^{+}, x \rangle} \ 1^{+} \wedge \qquad \frac{\Lambda, \langle A, 0^{+}, x \rangle \quad \Lambda, \langle B, 0^{+}, x \rangle}{\Lambda, \langle (A \wedge B), 0^{+}, x \rangle} \ 0^{+} \wedge$$

$$\frac{\Lambda, \langle A, 0^{-}, x \rangle, \langle B, 0^{-}, x \rangle}{\Lambda, \langle (A \wedge B), 0^{-}, x \rangle} \ 0^{-} \wedge \qquad \frac{\Lambda, \langle A, 1^{-}, x \rangle \quad \Lambda, \langle B, 1^{-}, x \rangle}{\Lambda, \langle (A \wedge B), 1^{-}, x \rangle} \ 1^{-} \wedge$$

$$\frac{\Lambda, \langle A, 1^{+}, x \rangle \quad \Lambda, \langle B, 1^{+}, x \rangle}{\Lambda, \langle (A \vee B), 1^{+}, x \rangle} \ 1^{+} \vee \qquad \frac{\Lambda, \langle A, 0^{+}, x \rangle, \langle B, 0^{+}, x \rangle}{\Lambda, \langle (A \vee B), 0^{+}, x \rangle} \ 0^{+} \vee$$

$$\frac{\Lambda, \langle A, 0^{-}, x \rangle \quad \Lambda, \langle B, 0^{-}, x \rangle}{\Lambda, \langle (A \vee B), 0^{-}, x \rangle} \ 0^{-} \vee \qquad \frac{\Lambda, \langle A, 1^{-}, x \rangle, \langle B, 1^{-}, x \rangle}{\Lambda, \langle (A \vee B), 1^{-}, x \rangle} \ 1^{-} \vee$$

$$\frac{\Lambda, \langle \Box_{\mathrm{F}} A, 1^{+}, x \rangle, \langle x, y \rangle, \langle A, 1^{+}, y \rangle}{\Lambda, \langle \Box_{\mathrm{F}} A, 1^{+}, x \rangle, \langle x, y \rangle} \ 1^{+} \Box_{\mathrm{F}} \qquad \frac{\Lambda, \langle A, 0^{+}, n \rangle, \langle x, n \rangle}{\Lambda, \langle \Box_{\mathrm{F}} A, 0^{+}, x \rangle} \ 0^{+} \Box_{\mathrm{F}}$$

$$\frac{\Lambda, \langle \Box_{\mathrm{F}} A, 0^{-}, x \rangle, \langle x, y \rangle, \langle A, 0^{-}, y \rangle}{\Lambda, \langle \Box_{\mathrm{F}} A, 0^{-}, x \rangle, \langle x, y \rangle} \ 0^{-} \Box_{\mathrm{F}} \qquad \frac{\Lambda, \langle A, 1^{-}, n \rangle, \langle x, n \rangle}{\Lambda, \langle \Box_{\mathrm{F}} A, 1^{-}, x \rangle} \ 1^{-} \Box_{\mathrm{F}}$$

$$\frac{\Lambda, \langle A, 1^{+}, n \rangle, \langle x, n \rangle}{\Lambda, \langle \Diamond_{\mathrm{F}} A, 1^{+}, x \rangle} \ 1^{+} \Diamond_{\mathrm{F}} \qquad \frac{\Lambda, \langle \Diamond_{\mathrm{F}} A, 0^{+}, x \rangle, \langle x, y \rangle, \langle A, 0^{+}, y \rangle}{\Lambda, \langle \Diamond_{\mathrm{F}} A, 0^{+}, x \rangle, \langle x, y \rangle} \ 0^{+} \Diamond_{\mathrm{F}}$$

$$\frac{\Lambda, \langle A, 0^{-}, n \rangle, \langle x, n \rangle}{\Lambda, \langle \Diamond_{\mathrm{F}} A, 0^{-}, x \rangle} \ 0^{-} \Diamond_{\mathrm{F}} \qquad \frac{\Lambda, \langle \Diamond_{\mathrm{F}} A, 1^{-}, x \rangle, \langle x, y \rangle, \langle A, 1^{-}, y \rangle}{\Lambda, \langle \Diamond_{\mathrm{F}} A, 1^{-}, x \rangle, \langle x, y \rangle} \ 1^{-} \Diamond_{\mathrm{F}}$$

$$\frac{\Lambda, \langle \Box_P A, \mathbf{1}^+, x\rangle, \langle y, x\rangle, \langle A, \mathbf{1}^+, y\rangle}{\Lambda, \langle \Box_P A, \mathbf{1}^+, x\rangle, \langle y, x\rangle} \; \mathbf{1}^+ \Box_P \qquad \frac{\Lambda, \langle A, \mathbf{0}^+, n\rangle, \langle n, x\rangle}{\Lambda, \langle \Box_P A, \mathbf{0}^+, x\rangle} \; \mathbf{0}^+ \Box_P$$

$$\frac{\Lambda, \langle \Box_P A, \mathbf{0}^-, x\rangle, \langle y, x\rangle, \langle A, \mathbf{0}^-, y\rangle}{\Lambda, \langle \Box_P A, \mathbf{0}^-, x\rangle, \langle y, x\rangle} \; \mathbf{0}^- \Box_P \qquad \frac{\Lambda, \langle A, \mathbf{1}^-, n\rangle, \langle n, x\rangle}{\Lambda, \langle \Box_P A, \mathbf{1}^-, x\rangle} \; \mathbf{1}^- \Box_P$$

$$\frac{\Lambda, \langle A, \mathbf{1}^+, n\rangle, \langle n, x\rangle}{\Lambda, \langle \Diamond_P A, \mathbf{1}^+, x\rangle} \; \mathbf{1}^+ \Diamond_P \qquad \frac{\Lambda, \langle \Diamond_P A, \mathbf{0}^+, x\rangle, \langle y, x\rangle, \langle A, \mathbf{0}^+, y\rangle}{\Lambda, \langle \Diamond_P A, \mathbf{0}^+, x\rangle, \langle y, x\rangle} \; \mathbf{0}^+ \Diamond_P$$

$$\frac{\Lambda, \langle A, \mathbf{0}^-, n\rangle, \langle n, x\rangle}{\Lambda, \langle \Diamond_P A, \mathbf{0}^-, x\rangle} \; \mathbf{0}^- \Diamond_P \qquad \frac{\Lambda, \langle \Diamond_P A, \mathbf{1}^-, x\rangle, \langle y, x\rangle, \langle A, \mathbf{1}^-, y\rangle}{\Lambda, \langle \Diamond_P A, \mathbf{1}^-, x\rangle, \langle y, x\rangle} \; \mathbf{1}^- \Diamond_P$$

$$\frac{\Lambda, \langle x, x\rangle}{\Lambda} \; T$$

$$\frac{\Lambda, \langle x, y\rangle, \langle y, z\rangle, \langle x, z\rangle}{\Lambda, \langle x, y\rangle, \langle y, z\rangle} \; 4$$

Remark 1 The natural number n must not appear in the conclusion of the rules: $\mathbf{0}^+ \Box_F$, $\mathbf{1}^- \Box_F$, $\mathbf{1}^+ \Diamond_F$, $\mathbf{0}^- \Diamond_F$, $\mathbf{0}^+ \Box_P$, $\mathbf{1}^- \Box_P$, $\mathbf{1}^+ \Diamond_P$, and $\mathbf{0}^- \Diamond_P$.

The notion of *derivation* as well as those of *initial sequent* and *endsequent* are defined inductively in the usual way. Roughly speaking, a derivation is a finite rooted tree in which the nodes are labelled sequents. The root of the tree (at the bottom) is called the endsequent and the leaves of the tree (at the top) are called initial sequents. The *length* of a derivation is the number of labelled sequents in that derivation.

Starting with the single set of rules of inference set out above, four notions of derivability are distinguished so that they differ only in the definition of axiomatic labelled sequent. A labelled sequent is *derivable*, *gap-derivable*, *glut-derivable*, or *classic-derivable* if there exists a derivation in which it is the endsequent and all initial sequents are respectively axiomatic, gap-axiomatic, glut-axiomatic, or classic-axiomatic.

Let Λ be a labelled sequent. Then:

- Λ is *axiomatic* if there is an atomic formula p and a natural number x such that either the labelled formulae $\langle p, \mathbf{1}^+, x\rangle$ and $\langle p, \mathbf{0}^+, x\rangle$ belong to Λ or the labelled formulae $\langle p, \mathbf{0}^-, x\rangle$ and $\langle p, \mathbf{1}^-, x\rangle$ belong to Λ.

- Λ is *gap-axiomatic* if it is axiomatic or there is an atomic formula p and a natural number x such that the labelled formulae $\langle p, \mathbf{1}^+, x\rangle$ and $\langle p, \mathbf{1}^-, x\rangle$ belong to Λ.

7

- Λ is *glut-axiomatic* if it is axiomatic or there is an atomic formula p and a natural number x such that the labelled formulae $\langle p, \mathbf{0}^-, x \rangle$ and $\langle p, \mathbf{0}^+, x \rangle$ belong to Λ.

- Λ is *classic-axiomatic* if it is gap-axiomatic or glut-axiomatic.

The general labelled sequent calculus is sound and complete with respect to the relational semantics. Moreover, these properties also hold for the gappy, glutty, and classical notions of validity and derivability.

Theorem 1 (soundness and completeness)   *Let Λ be a labelled sequent.*

1. *Λ is valid iff Λ is derivable.*

2. *Λ is gap-valid iff Λ is gap-derivable.*

3. *Λ is glut-valid iff Λ is glut-derivable.*

4. *Λ is classic-valid iff Λ is classic-derivable.*

Proof. The techniques for proving the soundness and completeness of such systems are well known. Also we refer the reader to (Priest, 2008) for a detailed proof of Theorem 1. Although the systems covered in the article do not involve past modalities, the proofs provided can be extended without difficulty to these two additional cases. □

3   Duality

The four-valued modal logic $\mathsf{K}_t^4\mathsf{T}4$ satisfies many duality properties. These properties rely on different types of symmetry. In this section, we point out three types of symmetry that are primitive and can be freely combined to define more complex forms of duality. The first two types correspond to an alethic symmetry (either qualitative or quantitative) while the third one corresponds to a relational symmetry. Also, we show properties associated with each of these types of symmetry. In this connection, the dual of a labelled sequent is specified in three different ways.

3.1   Two types of alethic duality

We distinguish a qualitative and a quantitative form of alethic duality. The qualitative alethic duality lies in the interchange of the maximum degree of truth and the maximum degree of falsehood on the one hand and the minimum degree of truth and the minimum degree of falsehood on the other hand. The quantitative alethic duality lies in the interchange of the maximum degree of truth and the minimum degree of truth on the one hand and the maximum degree of falsehood and the minimum degree of falsehood on the other hand. These two types of alethic duality emphasize a symmetry between logical connectives.

The *alethic dual* of a formula A of $\mathcal{L}(\mathsf{K_t})$, denoted by $[A]^\dagger$, is defined by induction on the complexity of A as follows:

$$
\begin{array}{rclcrcl}
[p]^\dagger & = & p & & [\Box_\mathsf{F} B]^\dagger & = & \Diamond_\mathsf{F}[B]^\dagger \\
[\neg B]^\dagger & = & \neg[B]^\dagger & & [\Diamond_\mathsf{F} B]^\dagger & = & \Box_\mathsf{F}[B]^\dagger \\
[(B \wedge C)]^\dagger & = & ([B]^\dagger \vee [C]^\dagger) & & [\Box_\mathsf{P} B]^\dagger & = & \Diamond_\mathsf{P}[B]^\dagger \\
[(B \vee C)]^\dagger & = & ([B]^\dagger \wedge [C]^\dagger) & & [\Diamond_\mathsf{P} B]^\dagger & = & \Box_\mathsf{P}[B]^\dagger
\end{array}
$$

3.1.1 Qualitative alethic duality

The qualitative alethic duality consists in swapping the degree of truth and the degree of falsehood. Proposition 3 states that none of the four notions of derivability are sensitive to this qualitative inversion.

The *qualitative alethic dual* of a labelled sequent Λ, denoted by $[\Lambda]^{QL}$, is the set $\{\langle [A]^\dagger, \overline{\lambda}, x \rangle | \langle A, \lambda, x \rangle \in \Lambda\} \cup \{\langle x, y \rangle | \langle x, y \rangle \in \Lambda\}$ where $\overline{\lambda}$ is defined as follows:

$$
\overline{\lambda} = \begin{cases}
\mathbf{0^+} & \text{if } \lambda = \mathbf{0^-} \\
\mathbf{1^-} & \text{if } \lambda = \mathbf{1^+} \\
\mathbf{0^-} & \text{if } \lambda = \mathbf{0^+} \\
\mathbf{1^+} & \text{if } \lambda = \mathbf{1^-}
\end{cases}
$$

Proposition 3 *Let Λ be a labelled sequent.*

1. *Λ is derivable iff $[\Lambda]^{QL}$ is derivable.*

2. *Λ is gap-derivable iff $[\Lambda]^{QL}$ is gap-derivable.*

3. *Λ is glut-derivable iff $[\Lambda]^{QL}$ is glut-derivable.*

4. *Λ is classic-derivable iff $[\Lambda]^{QL}$ is classic-derivable.*

Proof. By induction on the length of derivations. □

Vincent Degauquier

3.1.2 Quantitative alethic duality

The quantitative alethic duality consists in reversing the degree of truth on the one hand and the degree of falsehood on the other hand. Proposition 4 states that the properties of derivability and classic-derivability are not sensitive to this quantitative inversion while the properties of gap-derivability and glut-derivability are exchanged according to whether we are dealing with a labelled sequent or its quantitative alethic dual.

The *quantitative alethic dual* of a labelled sequent Λ, denoted by $[\Lambda]^{QT}$, is the set $\{\langle [A]^\dagger, \overline{\lambda}, x\rangle | \langle A, \lambda, x\rangle \in \Lambda\} \cup \{\langle x, y\rangle | \langle x, y\rangle \in \Lambda\}$ where $\overline{\lambda}$ is defined as follows:

$$\overline{\lambda} = \begin{cases} \mathbf{1}^- & \text{if } \lambda = \mathbf{0}^- \\ \mathbf{0}^+ & \text{if } \lambda = \mathbf{1}^+ \\ \mathbf{1}^+ & \text{if } \lambda = \mathbf{0}^+ \\ \mathbf{0}^- & \text{if } \lambda = \mathbf{1}^- \end{cases}$$

Proposition 4 *Let Λ be a labelled sequent.*

1. *Λ is derivable iff $[\Lambda]^{QT}$ is derivable.*

2. *Λ is gap-derivable iff $[\Lambda]^{QT}$ is glut-derivable.*

3. *Λ is glut-derivable iff $[\Lambda]^{QT}$ is gap-derivable.*

4. *Λ is classic-derivable iff $[\Lambda]^{QT}$ is classic-derivable.*

Proof. By induction on the length of derivations. □

3.1.3 Duality à la J. Michael Dunn

The combinaison of the qualitative alethic duality and the quantitative alethic duality gives rise to a form of duality pointed out by Dunn (1976) in the context of 'first degree entailment' and based on a symmetry between truth and non-falsehood on the one hand and falsehood and non-truth on the other hand. This observation is reflected in Proposition 5.

The *Dunn dual* of a labelled sequent Λ, denoted by $[\Lambda]^D$, is the set $\{\langle A, \overline{\lambda}, x\rangle | \langle A, \lambda, x\rangle \in \Lambda\} \cup \{\langle x, y\rangle | \langle x, y\rangle \in \Lambda\}$ where $\overline{\lambda}$ is defined as follows:

$$\overline{\lambda} = \begin{cases} 1^+ & \text{if } \lambda = 0^- \\ 0^- & \text{if } \lambda = 1^+ \\ 1^- & \text{if } \lambda = 0^+ \\ 0^+ & \text{if } \lambda = 1^- \end{cases}$$

Proposition 5 *For any Λ, $[\Lambda]^D = [[\Lambda]^{QL}]^{QT} = [[\Lambda]^{QT}]^{QL}$.*

3.2 Relational duality

The relational duality consists in swapping the future and past modalities while reversing the accessibility relation (see Burgess, 1984). Proposition 6 shows that the derivability, gap-derivability, glut-derivability, and classic-derivability of labelled sequents are not sensitive to this relational inversion. This is due to the fact that the accessibility relation with respect to the past is the inverse of the accessibility relation with respect to the future. In this sense, the future and past structures are the mirror of each other.

The *relational dual* of a formula A of $\mathcal{L}(\mathsf{K_t})$, denoted by $[A]^\ddagger$, is defined by induction on the complexity of A as follows:

$$
\begin{aligned}
[p]^\ddagger &= p & [\Box_\mathsf{F} B]^\ddagger &= \Box_\mathsf{P}[B]^\ddagger \\
[\neg B]^\ddagger &= \neg[B]^\ddagger & [\Diamond_\mathsf{F} B]^\ddagger &= \Diamond_\mathsf{P}[B]^\ddagger \\
[(B \wedge C)]^\ddagger &= ([B]^\ddagger \wedge [C]^\ddagger) & [\Box_\mathsf{P} B]^\ddagger &= \Box_\mathsf{F}[B]^\ddagger \\
[(B \vee C)]^\ddagger &= ([B]^\ddagger \vee [C]^\ddagger) & [\Diamond_\mathsf{P} B]^\ddagger &= \Diamond_\mathsf{F}[B]^\ddagger
\end{aligned}
$$

The *relational dual* of a labelled sequent Λ, denoted by $[\Lambda]^R$, is the set $\{\langle [A]^\ddagger, \lambda, x \rangle | \langle A, \lambda, x \rangle \in \Lambda\} \cup \{\langle x, y \rangle | \langle y, x \rangle \in \Lambda\}$.

Proposition 6 *Let Λ be a labelled sequent.*

1. Λ is derivable iff $[\Lambda]^R$ is derivable.

2. Λ is gap-derivable iff $[\Lambda]^R$ is gap-derivable.

3. Λ is glut-derivable iff $[\Lambda]^R$ is glut-derivable.

4. Λ is classic-derivable iff $[\Lambda]^R$ is classic-derivable.

Proof. By induction on the length of derivations. $\qquad\qquad\square$

4 Unifying partiality and paraconsistency

Several well-known partial and paraconsistent logics can be faithfully embedded into $\mathsf{K}_t^4\mathsf{T4}$. In this section, three many-valued logics and three constructive logics are addressed. Among the many-valued logics, we consider Kleene's strong three-valued logic ($\mathsf{K_3}$), Priest's logic of paradox (LP), and Dunn-Belnap's four-valued logic ($\mathsf{L_4}$). As far as the constructive logics are concerned, intuitionistic logic (H), dual-intuitionistic logic (B), and bi-intuitionistic logic (HB) are investigated (see Drobyshevich, Odintsov, & Wansing, 2022; Goré, 2000; Rauszer, 1980).

In order to state these embeddings precisely, some definitions are needed. A *sequent* for a language $\mathcal{L}$ is an ordered pair $\langle \Gamma, \Delta \rangle$, where Γ and Δ are finite sequences of formulae of $\mathcal{L}$. The sequent $\langle \Gamma, \Delta \rangle$ is denoted $\Gamma \vdash \Delta$ and is said to be L-valid if it is valid according to the logic L. Moreover, if Σ is a sequence of formulae $A_1, \ldots, A_n$, then $\langle \Sigma, \lambda, x \rangle$ denotes the labelled sequent $\{\langle A_i, \lambda, x \rangle | 1 \leq i \leq n \}$.

4.1 Embedding many-valued logics

The language of the many-valued logics with which we are concerned is the language of $\mathsf{K}_t^4\mathsf{T4}$ without modal symbols. This language, here referred to as $\mathcal{L}(\mathsf{CL})$, is actually the language of classical propositional logic (CL). It is to be noted that such a language usually includes an additional logical symbol interpreted as material implication. In this context, this symbol is denoted $\supset$ and a formula of the form $(A \supset B)$ is regarded as an abbreviation of $(\neg A \vee B)$.

Proposition 7 *Let $\Gamma \vdash \Delta$ be a sequent for $\mathcal{L}(\mathsf{CL})$.*

1. *$\Gamma \vdash \Delta$ is $\mathsf{L_4}$-valid iff $\langle \Gamma, \mathbf{1}^+, x \rangle, \langle \Delta, \mathbf{0}^+, x \rangle$ is derivable.*

2. *$\Gamma \vdash \Delta$ is $\mathsf{K_3}$-valid iff $\langle \Gamma, \mathbf{1}^+, x \rangle, \langle \Delta, \mathbf{0}^+, x \rangle$ is gap-derivable.*

3. *$\Gamma \vdash \Delta$ is LP-valid iff $\langle \Gamma, \mathbf{1}^+, x \rangle, \langle \Delta, \mathbf{0}^+, x \rangle$ is glut-derivable.*

4. *$\Gamma \vdash \Delta$ is CL-valid iff $\langle \Gamma, \mathbf{1}^+, x \rangle, \langle \Delta, \mathbf{0}^+, x \rangle$ is classic-derivable.*

Proof. This results from Theorem 1 and the fact that $\Gamma \vdash \Delta$ is respectively $\mathsf{L_4}$-valid, $\mathsf{K_3}$-valid, LP-valid, and CL-valid if and only if $\langle \Gamma, \mathbf{1}^+, x \rangle, \langle \Delta, \mathbf{0}^+, x \rangle$ is valid, gap-valid, glut-valid, and classic-valid. For each of the many-valued logics discussed, we need to specify the notions of model, truth, falsehood, and validity. It then remains to establish by induction on the complexity of formulae that there is a counter-model, a consistent counter-model, a complete counter-model, and a classical counter-model to $\langle \Gamma, 1^+, x \rangle, \langle \Delta, 0^+, x \rangle$ if and only if there is a counter-model to $\Gamma \vdash \Delta$ for L_4, K_3, LP, and CL, respectively. □

4.2   Embedding constructive logics

Intuitionistic, dual-intuitionistic, and bi-intuitionistic propositional logics each have a different language. The language of intuitionistic logic, denoted by $\mathcal{L}(H)$, has $\sim$, $\cap$, $\cup$ and $\to$ as logical symbols. The language of dual-intuitionistic logic, denoted by $\mathcal{L}(B)$, has $-$, $\cap$, $\cup$ and $\prec$ as logical symbols. Finally, bi-intuitionistic logic involves all of these logical symbols and its language is denoted by $\mathcal{L}(HB)$. As for the syntax of these three languages, the formulae are defined inductively in the usual way.

Let us define a translation function τ (see Łukowski, 1996) from the set of formulae of $\mathcal{L}(HB)$ to the set of formulae of $\mathcal{L}(K_t)$. This translation consists in an extension of the translation proposed by McKinsey and Tarski (1948) and is defined by the following clauses:

$$\begin{aligned}
\tau[p] &= \Box_F\, p \text{ or } \Diamond_P\, p \\
\tau[\sim A] &= \Box_F\, \neg\tau[A] \\
\tau[-A] &= \Diamond_P\, \neg\tau[A] \\
\tau[(A \cap B)] &= (\tau[A] \wedge \tau[B]) \\
\tau[(A \cup B)] &= (\tau[A] \vee \tau[B]) \\
\tau[(A \to B)] &= \Box_F(\neg\tau[A] \vee \tau[B]) \\
\tau[(A \prec B)] &= \Diamond_P(\neg\tau[B] \wedge \tau[A])
\end{aligned}$$

To simplify the notation, we adopt the convention that if Σ is a sequence of formulae $A_1, \ldots, A_n$, then $\tau[\Sigma]$ denotes the sequence $\tau[A_1], \ldots, \tau[A_n]$.

Proposition 8   *Let $\Gamma \vdash \Delta$ be a sequent for $\mathcal{L}(H)$. Then, $\Gamma \vdash \Delta$ is H-valid if and only if $\langle \tau[\Gamma], 1^+, x \rangle, \langle \tau[\Delta], 0^+, x \rangle$ is classic-derivable.*

Proposition 9   *Let $\Gamma \vdash \Delta$ be a sequent for $\mathcal{L}(B)$. Then, $\Gamma \vdash \Delta$ is B-valid if and only if $\langle \tau[\Gamma], 1^+, x \rangle, \langle \tau[\Delta], 0^+, x \rangle$ is classic-derivable.*

Proposition 10   *Let $\Gamma \vdash \Delta$ be a sequent for $\mathcal{L}(HB)$. Then, $\Gamma \vdash \Delta$ is HB-valid if and only if $\langle \tau[\Gamma], 1^+, x \rangle, \langle \tau[\Delta], 0^+, x \rangle$ is classic-derivable.*

Proof. As Propositions 8–9 are special cases of Proposition 10, we only sketch the proof of the latter. By Theorem 1, it suffices to show that $\Gamma \vdash \Delta$ is HB-valid if and only if $\langle \tau[\Gamma], 1^+, x \rangle, \langle \tau[\Delta], 0^+, x \rangle$ is classic-valid.

Vincent Degauquier

A bi-intuitionistic model for the language $\mathcal{L}(\mathsf{HB})$ is defined as a classical model $\mathcal{M} = \langle W, R, V \rangle$ which satisfies the following persistence condition: for all α and β in W, if $\langle \alpha, \beta \rangle \in R_\mathsf{F}$, then $\alpha \in V^+(n)$ implies $\beta \in V^+(n)$ (for every $n \in \mathbb{N}$). The bi-intuitionistic truth (denoted by $\mathcal{M}, \alpha \vDash$) and falsehood (denoted by $\mathcal{M}, \alpha \nvDash$) of the formulae of $\mathcal{L}(\mathsf{HB})$ as well as the HB-validity are defined as usual (see Goré, 2000).

To prove that a sequent $\Gamma \vdash \Delta$ is not HB-valid if and only if the labelled sequent $\langle \tau[\Gamma], \mathbf{1}^+, x \rangle, \langle \tau[\Delta], \mathbf{0}^+, x \rangle$ is not classic-valid, we establish that, for all classical models $\mathcal{M}$ and for all formulae A of $\mathcal{L}(\mathsf{HB})$, $\mathcal{M}, \alpha \vDash^+ \tau[A]$ if and only if $\mathcal{M}$ satisfies the persistence condition and $\mathcal{M}, \alpha \vDash A$. This is done by induction on the complexity of formulae. □

4.3 A general principle of duality

To propose a unified view of partiality and paraconsistency, a general principle of duality resulting from the combination of the quantitative alethic duality and the relational duality is identified and shown to apply to every partial or paraconsistent logic discussed. Through this general principle and their embedding into $\mathsf{K}_t^4\mathsf{T}4$, a perfect symmetry is observed between Kleene's strong three-valued logic and Priest's logic of paradox on the one hand and between intuitionistic logic and dual-intuitionistic logic on the other hand. In addition, Dunn-Belnap's four-valued logic, just like bi-intuitionistic logic, is its own counterpart.

The *general dual* of a labelled sequent Λ, denoted by $\delta[\Lambda]$, is defined as $[[\Lambda]^{QT}]^R$ or, equivalently, $[[\Lambda]^R]^{QT}$.

Proposition 11 *Let $\Gamma \vdash \Delta$ and $\Gamma' \vdash \Delta'$ be sequents for $\mathcal{L}(\mathsf{CL})$ such that $\langle \Gamma, \mathbf{1}^+, x \rangle, \langle \Delta, \mathbf{0}^+, x \rangle = \delta[\langle \Gamma', \mathbf{1}^+, x \rangle, \langle \Delta', \mathbf{0}^+, x \rangle]$.*

1. $\Gamma \vdash \Delta$ is CL-valid iff $\Gamma' \vdash \Delta'$ is CL-valid.

2. $\Gamma \vdash \Delta$ is K_3-valid iff $\Gamma' \vdash \Delta'$ is LP-valid.

3. $\Gamma \vdash \Delta$ is L_4-valid iff $\Gamma' \vdash \Delta'$ is L_4-valid.

Proposition 12 *For any $\Gamma \vdash \Delta$ for $\mathcal{L}(\mathsf{H})$ and $\Gamma' \vdash \Delta'$ for $\mathcal{L}(\mathsf{B})$, if $\langle \tau[\Gamma], \mathbf{1}^+, x \rangle, \langle \tau[\Delta], \mathbf{0}^+, x \rangle = \delta[\langle \tau[\Gamma'], \mathbf{1}^+, x \rangle, \langle \tau[\Delta'], \mathbf{0}^+, x \rangle]$, then $\Gamma \vdash \Delta$ is H-valid if and only if $\Gamma' \vdash \Delta'$ is B-valid.*

Proposition 13 *Let $\Gamma \vdash \Delta$ and $\Gamma' \vdash \Delta'$ be sequents for $\mathcal{L}(\mathsf{HB})$ such that $\langle \tau[\Gamma], \mathbf{1}^+, x \rangle, \langle \tau[\Delta], \mathbf{0}^+, x \rangle = \delta[\langle \tau[\Gamma'], \mathbf{1}^+, x \rangle, \langle \tau[\Delta'], \mathbf{0}^+, x \rangle]$. Then, $\Gamma \vdash \Delta$ is HB-valid if and only if $\Gamma' \vdash \Delta'$ is HB-valid.*

5 Conclusion

In light of the foregoing results, it appears that while several logico-philosophical meanings of partiality and paraconsistency can be distinguished, they all obey the same principle of duality. The composition of the quantitative alethic duality function and the relational duality function defined on the set of labelled sequents gives rise to a general principle of duality applying to every partial or paraconsistent logic discussed. Among many-valued logics, Kleene's strong three-valued logic and Priest's logic of paradox are dual of each other and Dunn-Belnap's four-valued logic is self-dual. Among constructive logics, intuitionistic logic and dual-intuitionistic logic are dual of each other and bi-intuitionistic logic is its own dual.

We plan to extend the results obtained in this article in two directions. First, we intend to draw up an exhaustive list of the principles of duality existing in the literature and to investigate whether these principles can be defined on the basis of the three types of duality we have selected (see, for example, Brunner & Carnielli, 2005; Dunn, 2000; Wansing, 2010). In particular, an analysis of the bi-intuitionistic logic 2Int developed in (Wansing, 2013) would be of great interest to our study of different forms of duality. While this logic is closely related to that proposed in (Rauszer, 1980), it relies on a different principle of duality (see Drobyshevich, 2019). Also, it would be worthwhile to determine to what extent the concept of 'Gentzen dual' presented in (Kracht, 1996) would correspond to a principle of duality resulting from the combination of the principle of relational duality and one of the two principles of alethic duality. Second, we intend to examine whether our general principle of duality also applies to non-classical logics involving other notions of partiality and paraconsistency. Among the well-known non-classical logics that we plan to discuss are Łukasiewicz's three-valued logic, Gödel's three-valued logic, Johansson's minimal logic and Nelson's four-valued constructive logic, which can be faithfully embedded into positive intuitionistic logic (see Kamide & Wansing, 2015).

References

Belnap, N. D. (1977). A useful four-valued logic. In J. M. Dunn & G. Epstein (Eds.), *Modern uses of multiple-valued logic* (pp. 8–37). Dordrecht: Reidel Publishing Company.

Bochman, A. (1998). Biconsequence relations: a four-valued formalism of reasoning with inconsistency and incompleteness. *Notre Dame Journal of Formal Logic*, *39*(1), 47–73.

Bonnette, N., & Goré, R. (1998). A labelled sequent system for tense logic K_t. In G. Antoniou & J. Slaney (Eds.), *Advanced topics in artificial intelligence. 11th Australian joint conference on artificial intelligence, AI'98. Brisbane, Australia, July 13–17, 1998. Selected papers* (pp. 71–82). Berlin: Springer-Verlag.

Brunner, A. B. M., & Carnielli, W. A. (2005). Anti-intuitionism and paraconsistency. *Journal of Applied Logic*, *3*(1), 161–184.

Burgess, J. P. (1984). Basic tense logic. In D. Gabbay & F. Guenthner (Eds.), *Handbook of philosophical logic. Volume II. Extensions of classical logic* (pp. 89–133). Dordrecht: Reidel Publishing Company.

Degauquier, V. (2018). A useful four-valued extension of the temporal logic $K_t T4$. *Bulletin of the Section of Logic*, *47*(1), 15–31.

Drobyshevich, S. (2019). A bilateral hilbert-style investigation of 2-intuitionistic logic. *Journal of Logic and Computation*, *29*(5), 665–692.

Drobyshevich, S., Odintsov, S., & Wansing, H. (2022). Moisil's modal logic and related systems. In K. Bimbó (Ed.), *Relevance logics and other tools for reasoning. Essays in honor of J. Michael Dunn* (pp. 150–177). London: College Publications.

Dunn, J. M. (1976). Intuitive semantics for first-degree entailments and 'coupled trees'. *Philosophical Studies*, *29*(3), 149–168.

Dunn, J. M. (2000). Partiality and its dual. *Studia Logica*, *66*(1), 5–40.

Girard, J.-Y. (1976). Three-valued logic and cut-elimination: the actual meaning of Takeuti's conjecture. *Dissertationes Mathematicae (Rozprawy Matematyczne)*, *136*, 1–49.

Gödel, K. (1986). On the intuitionistic propositional calculus. In S. Feferman (Ed.), *Collected works. Volume I. Publications 1929–1936* (pp. 222–225). New York: Oxford University Press.

Goré, R. (2000). Dual intuitionistic logic revisited. In R. Dyckhoff (Ed.), *Automated reasoning with analytic tableaux and related methods. International conference, TABLEAUX 2000. St Andrews, Scotland, UK, July 3–7, 2000. Proceedings* (pp. 252–267). Berlin: Springer-Verlag.

Kamide, N., & Wansing, H. (2015). *Proof theory of N4-related paraconsistent logics*. London: College Publications.

Kracht, M. (1996). Power and weakness of the modal display calculus. In H. Wansing (Ed.), *Proof theory of modal logic* (pp. 93–122). Dordrecht: Springer.

Łukowski, P. (1996). Modal interpretation of Heyting-Brouwer logic. *Bulletin of the Section of Logic*, *25*(2), 80–83.

McKinsey, J. C. C., & Tarski, A. (1948). Some theorems about the sentential calculi of Lewis and Heyting. *The Journal of Symbolic Logic*, *13*(1), 1–15.

Muskens, R. (1999). On partial and paraconsistent logics. *Notre Dame Journal of Formal Logic*, *40*(3), 352–374.

Negri, S. (2005). Proof analysis in modal logic. *Journal of Philosophical Logic*, *34*(5/6), 507–544.

Priest, G. (2008). Many-valued modal logics: a simple approach. *The Review of Symbolic Logic*, *1*(2), 190–203.

Rauszer, C. (1980). An algebraic and Kripke-style approach to a certain extension of intuitionistic logic. *Dissertationes Mathematicae (Rozprawy Matematyczne)*, *167*, 1–62.

Rescher, N., & Urquhart, A. (1971). *Temporal logic*. Wien: Springer-Verlag.

Restall, G. (2004). Laws of non-contradiction, laws of the excluded middle, and logics. In G. Priest, J. Beall, & B. Armour-Garb (Eds.), *The law of non-contradiction. New philosophical essays* (pp. 73–84). Oxford: Clarendon Press.

Wansing, H. (2010). Proofs, disproofs, and their duals. In L. Beklemishev, V. Goranko, & V. Shehtman (Eds.), *Advances in modal logic. Volume 8* (pp. 483–505). London: College Publications.

Wansing, H. (2013). Falsification, natural deduction and bi-intuitionistic logic. *Journal of Logic and Computation*, *26*(1), 425–450.

Vincent Degauquier
University of Namur, Faculty of Sciences
Belgium
E-mail: vincent.degauquier@gmail.com

Subject-Matter and Intensional Operators IV: Left and Right Topic Sufficiency

Thomas Macaulay Ferguson[1]

Abstract: Topic-sensitive logical frameworks like William Parry's PAI or Francesco Berto's topic-sensitive intentional modals tend to share an assumption that intensional connectives are topic-transparent, *i.e.*, do not contribute or alter the topics of subformulae. Recent work in the setting of Parry has introduced logics whose implicit theory of topic allows for intensional operators to be transformative, including a system S/PAI in which the topic-theoretic contribution of the antecedent and consequent is not restricted to their topics alone, but the states to which they refer. The degree to which the antecedent and consequent contribute state-sensitive subject-matter varies with context; in this paper, we consider extensions of S/PAI in which the contribution of one or the other is limited to its topic alone.

Keywords: subject-matter, analytic implication, theory of topic

1 Introduction

Topic-sensitive logical frameworks like William Parry's PAI of (Parry, 1968) or Francesco Berto's topic-sensitive intentional modals of (Berto, 2022) tend to share an assumption that intensional connectives are topic-transparent, *i.e.*, do not contribute or alter the topics of subformulae. Recently, more fine-grained models for the subject-matter of intensional sentences have been developed over a sequence of papers (Ferguson, 2021, 2023a, 2023b, 2023c).

The logic S/PAI of (Ferguson, 2023c) allows for *state-sensitive subject-matter* that corresponds to an intuition that the states over which an intensional conditional is evaluated constitute part of the overall topic of the

---

[1]I am grateful for the participants of Logica 2022 for a great deal of fantastic feedback. Nicholas Ferenz's comments were especially impactful, both formally and philosophically. This paper is an outcome of the project Logical Structure of Information Channels, no. 21-23610M, supported by the Czech Science Foundation and carried out at the Institute of Philosophy of the Czech Academy of Sciences.

conditional. The degree to which the antecedent or consequent make such a contribution is determined by context, motivating the exploration of extensions of the system in which the preeminence of one over the other is respected.

2   From Intensional Transparency to State Sensitivity

An intuition that intensional operators play a *transformational* role modifying the topic of subformulae is natural, as *e.g.* reflected in Berto's remarks:

> [I]t's uncertain whether all the vocabulary we may want to call 'logical' is topic-transparent. Surely the topic-sensitive intentional operators... are not... 'Necessarily, John is human' seems to address a different topic from 'John is human' in a number of natural conversational contexts. (Berto, 2022, p. 34)

As we work in the context of Parry-style logics, we focus on analytic implication as a proxy for any intensional connective.[2] For an intensional conditional →, a thesis of *Intensional Transparency* that intensional connectives are topic-transparent, where *t* assigns topics and ⊕ is simple topic fusion, can be expressed:

Intensional Transparency $t(\varphi \to \psi) = t(\varphi) \oplus t(\psi)$

This constraint is implicit in both Fine's semantics for PAI in (Fine, 1986) and Berto's treatment of topic-sensitive intentional modals in (Berto, 2022).

2.1   Counterexamples to Intensional Transparency

In (Ferguson, 2023a), a number of reasons to question Intensional Transparency were explored, ultimately driving the shape of the requirements by which the logic CA/PAI was introduced. We can review two of the general reasons to question Intensional Transparency in the case of an intensional conditional to mark out the first leg of the path to state sensitivity.

A first consequence of Intensional Transparency is that intensional conditionals will be required to be at best *explicative*, that is, the condition demands that a conditional's overall subject-matter cannot exceed the joint subject-matters of its subformulae. Formally:

---

[2]But see (Ferguson, 2023b) for the type of analyses of unary modal operators that Berto suggests is needed.

Explicativity of Conditionals $t(\varphi \rightarrow \psi) \leq t(\varphi) \oplus t(\psi)$

There are a number of intensional contexts for which such explicativity requirements run counter to intuitions. An example given in (Ferguson, 2023a) and developed in (Ferguson, 2023b) is that of *intuitionistic implication*.

On the famed BHK reading, to assert a conditional $\varphi \rightarrow \psi$ is to attest to one's possessing a general method f of constructing proofs of ψ from proofs of φ. It is not unreasonable, on this reasoning, to say that $\varphi \rightarrow \psi$ is *about* this construction f. But this topic is not necessarily discoverable in the topics of φ and ψ alone; it is subject-matter that *exceeds* that of its parts. Arguably, in intuitionistic contexts, intensional connectives amplify topic beyond what is included within the simple fusion of the topics of their subsentences. Similar violations appear to hold of topic-sensitive intensional modals championed by Berto. *E.g.* on the knowability-relative-to-information interpretation of (Berto & Hawke, 2021), the intentional formula $\mathsf{K}^\varphi\psi$—read as "an agent is in a position to know that ψ given the information φ"—is in part about *knowability*, a topic that may be absent in the topics of φ and ψ.

The second consequence of Intensional Transparency to point out is that conditionals must at most be *ampliative*, requiring that the fusion of the topics of an antecedent and consequent are included (not necessarily properly) in the topic of a conditional. Put more precisely:

Ampliativity of Conditionals $t(\varphi) \oplus t(\psi) \leq t(\varphi \rightarrow \psi)$

The context of counterfactual conditionals is one in which such ampliativity runs counter to standard intuitions, witnessed by cases in which a conditional's topic omits some content from the topics of its subsentences.

A straightforward route to such cases appeals to the context-sensitivity with which definite descriptions refer to objects. Suppose $\psi(x)$ is a predicate corresponding to a contingent property that can be satisfied by at most one object and that the definite description $\imath x\psi(x)$ picks out an object a in the actual world. Let $\varphi(x)$ be a property such that $\varphi(a)$ is necessarily false although $\varphi(\imath x\psi(x))$ is possible. Then although $\varphi(\imath x\psi(x))$ is a sentence about a—even if it ascribes something *falsely* to a—a counterfactual conditional $\varphi(\imath x\psi(x)) \rightarrow \xi$ will only be evaluated in contexts in which the antecedent is true, *i.e.*, contexts in which the antecedent is *not* about a. To illustrate:

1 "Were the current Secretary-General of the United Nations a bonobo, the world would be much more peaceful."

As António Gutteres is the current Secretary-General, Gutteres inarguably constitutes part of the topic of the antecedent of [1]. But [1] is evaluated *strictly* with respect to counterfactual contexts in which the current Secretary-General is a bonobo; as Gutteres is metaphysically incapable of being a bonobo, this individual must be *omitted* from the overall subject-matter of [1]. There is thus an element of the fusion of the topics of the antecedent and consequent that is lacking in the topic of the counterfactual conditional.

2.2 Counterexamples to Topic Sufficiency

The system introduced in (Ferguson, 2023a) provided a framework whose underlying theory of topic avoids the foregoing counterintuitive features of *e.g.* Parry's PAI, among others. Nevertheless, as discussed in (Ferguson, 2023c), this did not eliminate *all* problematic features of its predecessors.

The new framework preserved PAI's implicit commitment to a condition that *the topic of an intensional conditional is a function of the subject-matters of its antecedent and consequent*. Informally, this condition can be read as a stipulation that the topic of a complex can be entirely determined by the topics of its parts. To make this characterization more precise, we can represent this formally as a principle of *Topic Sufficiency*:

- **Topic Sufficiency:** For an intensional conditional $\to$, if $t(\varphi) = t(\xi)$ and $t(\psi) = t(\zeta)$, then $t(\varphi \to \psi) = t(\xi \to \zeta)$.

The condition is remarkably entrenched, reflected in virtually all topic-theoretic frameworks in which conditionals are assigned topics. There exists phenomenological evidence against this principle, too, whence we will consider some informal examples showing why one might take issue with *Topic Sufficiency*.

For the first illustration, consider a scenario in which Simone is enrolled in two classes—biology and geography—and her friends are discussing a celebration for Simone in case she passes each of her classes. The conversation might be initiated as:

Q1 'How should we celebrate if Simone passes all of her classes?'

Again, consider two statements that might be offered in response to [Q2]:

R1 'Should Simone pass biology *and* geography, she would love a party.'
R2 'Should Simone pass biology *or* geography, she would love a party.'

Left and Right Topic Sufficiency

By *Junctive Transparency*—that extensional conjunction is topic-transparent in the sense of (Berto, 2022)—the antecedents of [R1] and [R2] share identical subject-matter, whence *Topic Sufficiency* predicts the coincidence of the topics of [R1] and [R2].

However, the question [Q1] identified those contingencies in which Simone passes biology *and* geography as the domain of topics germane to the ensuing discussion. In uttering [R1], a speaker talks about those hypothetical states in which Simone passes *both classes*, thereby remaining within the topic-theoretic bounds determined by [Q1]. In uttering [R2], the speaker talks about a *broader*—and thus *distinct*—collection of situations. While the topic of [R1] is appropriate to the discussion, that of [R2] fails to observe these constraints. One would therefore anticipate that the questioner would *reject* [R2] as *off-topic*.

The prediction that the subject-matters of [R1] and [R2] are identical requires that [R1] is on-topic precisely when [R2] is on-topic; a single topic cannot be both on-topic and off-topic with respect to a single context. The condition of *Topic Sufficiency* therefore conflicts with very plausible assumptions concerning the topics of intensional conditionals. To the extent that the subject-matters of the antecedents of [R1] and [R2] are insufficient to determine the antecedents' truth-conditions, topic of subsentences alone is not sufficient to pin down topics of an intensional complex.

Theories like Kratzer's (Kratzer, 1991)—in which a conditional's antecedent serves as a *restictor* constraining the worlds over which a conditional is evaluated—tend to emphasize the role of the antecedent in determining the stative subject-matter of a conditional. However there are cases in which the consequent plays a similar role in determining the states that a conditional is about. Ed Mares' (Mares, 2004, p. 144–146) recalls an example from Gabbay's (Gabbay, 1972) showing how consequents influence the class of states that a conditional is about:

2 'If I were the Pope, I would have allowed the use of the pill in India.'

3 'If I were the Pope, I would have dressed more humbly.'

Gabbay notes "clearly, in the first statement, we must assume that India remains overpopulated and poor in resources, while in the second example nothing of the sort is required."(Gabbay, 1972, p. 98) Such requirements flow from the consequent, however. For this reason, we must be able to account for cases in which the consequent is the primary subformula that pins down the context.

The degree to which the antecedent or consequent plays a role in fixing the states over which a conditional are evaluated varies with context. In *e.g.* a context like [Q1], the antecedent seems to eclipse the role of the consequent; in an abductive setting in which one is attempting to find plausible explanations for a consequent, the role of the antecedent is minimized. For this reason, we might be interested in systems in which one can distinguish *left topic sufficiency* (when an antecedent's contribution is its topic alone) from *right topic sufficiency* (when the consequent's contribution is restricted to its topic). In what follows, we will review the system of state-sensitive analytic implication (S/PAI) of (Ferguson, 2023c) and examine the semantic and syntactic modifications necessary to capture extensions satisfying left or right topic sufficiency.

3   The Logic of State-Sensitive Analytic Implication

In this section, we will review the lofic of state-sensitive analytic implication (S/PAI) introduced in (Ferguson, 2023c); this will be the setting in which we will examine left and right topic suffiency. Fix a propositional language $\mathcal{L}$ including negation ($\neg$), conjunction ($\wedge$), disjunction ($\vee$), a material conditional ($\supset$), and an intensional "analytic implication" conditional ($\rightarrow$).[3]

3.1   Models for State-Sensitive Analytic Implication

We will first introduce S/PAI model-theoretically by defining *state-sensitive Fine models*. For notation, we let $w\uparrow$ denote the *R*-cone of *w*. Then:

Definition 1   *An* S/PAI *Fine model is a tuple* $\langle W, R, \mathcal{T}, C, \oplus, \multimap, v, t \rangle$ *such that:*

- *For each $w \in W$, $\langle \mathcal{T}_w, \oplus_w \rangle$ is a join semilattice*
- *v is a valuation from atomic formulae to W*
- *For each $w \in W$, t_w is a function mapping atomic formulae to $\mathcal{T}_w$*
- *For each $w \in W$, C_w is a set of* contents *defined recursively below*
- *For any $\langle a, X \rangle \in C_w$, $h_{w,w'} = \langle h_{w,w'}(a), X \cap (w\uparrow) \rangle$*
- *For each $w \in W$, $\multimap_w$ is a binary function from $C_w \times C_w \rightarrow \mathcal{T}_w$*
- *For all w, w' such that wRw', $h_{w,w'}$ is a function such that:*

    - *for atoms p, $h_{w,w'}(t_w(p)) = t_{w'}(p)$*

---

[3]For the sake of space, we treat $\vee$ and $\supset$ as definable from $\neg$ and $\wedge$ in the usual fashion.

- $h_{w,w'}(a \oplus_w b) = h_{w,w'}(a) \oplus_{w'} h_{w,w'}(b)$ for $a, b \in \mathcal{T}_w$
- $h_{w,w'}(c -\!\!\circ_w d) = h_{w,w'}(c) -\!\!\circ_{w'} h_{w,w'}(d)$ for $c, d \in C_w$

Note that each $\langle \mathcal{T}_w, \oplus_w \rangle$ induces a partial order $\leq_w$ by setting $a \leq_w b$ if $a \oplus_w b = b$. As each function $h_{w,w'}$ preserves $\oplus_w$, the order will also be respected by $h_{w,w'}$. The introduction of such functions ensures the preservation of topic inclusion along R. The $-\!\!\circ$ function guides the assignment of topics and contents to intentional formulae of the form $\varphi \to \psi$. In contrast to (Ferguson, 2023a), the functions $h_{w,w'}$ are *polymorphic*, taking both topics from $\mathcal{T}_w$ and contents from C_w as arguments. This calls for more precisely defining contents—*i.e.*, the members of set C_w—by recursively defining a function c_w assigning contents to sentences. This must be done in coordination with truth-conditions, so the following two definitions are invoked in a double recursive definition. Let $(\!|\varphi|\!)_w$ denote the set $\{w' \in w{\uparrow} \mid w' \Vdash \varphi\}$.

Definition 2   *We define a function c_w. Where π_0 and π_1 are projection functions onto first and second coordinates, respectively, let $c_w^0 = \pi_0 \circ c_w$ and $c_w^1 = \pi_1 \circ c_w$. Then:*

- $c_w(p) = \langle t_w(p), v(p) \cap w{\uparrow} \rangle$
- $c_w(\neg\varphi) = \langle c_w^0(\varphi), w{\uparrow} \setminus c_w^1(\varphi) \rangle$
- $c_w(\varphi \wedge \psi) = \langle c_w^0(\varphi) \oplus_w c_w^0(\psi), c_w^1(\varphi) \cap c_w^1(\psi) \rangle$
- $c_w(\varphi \to \psi) = \langle c_w(\varphi) -\!\!\circ_w c_w(\psi), (\!|\varphi \to \psi|\!)_w \rangle$

The first coordinate serves as the topic, whence we define $t_w(\varphi)$ as $c_w^0(\varphi)$.

In tandem with this definition, we define truth at a world:

Definition 3   *Truth conditions are defined recursively:*

- $w \Vdash p$ *if* $w \in v(p)$
- $w \Vdash \neg\varphi$ *if* $w \nVdash \varphi$
- $w \Vdash \varphi \wedge \psi$ *if* $w \Vdash \varphi$ *and* $w \Vdash \psi$
- $w \Vdash \varphi \to \psi$ *if* $\begin{cases} \text{for all } w' \text{ such that } wRw', \text{ if } w' \Vdash \varphi \text{ then } w' \Vdash \psi \\ t_w(\psi) \leq_w t_w(\varphi) \end{cases}$

c_w induces a definition for each C_w:

Definition 4   *The set of contents C_w at a world w is the set $c_w[\mathcal{L}]$—the image of the language under c_w*

Consequence can be defined as usual.

3.2 Axioms for State Sensitive Analytic Implication

We can now provide a Hilbert-style calculus for S/PAI. Two notational comments are in order. First, if $f(\varphi)$ is a formula in which φ appears, $f(\psi)$ is a formula resulting from the replacement of one or more instances of φ with ψ in $f(\varphi)$. Second, we use the notation $\mathbf{t}_\varphi$ as shorthand for the formula $\varphi \supset \varphi$ (this is, essentially, a signed version of the Ackermann $\mathbf{t}$). With these notational issues covered, we describe an axiomatization of S/PAI:

Definition 5 *The logic of state-sensitive analytic implication S/PAI is determined by the following axioms:*

[A1]	$(\varphi \wedge \psi) \rightarrow (\psi \wedge \varphi)$
[A2]	$\varphi \rightarrow (\varphi \wedge \varphi)$
[A3]	$\varphi \rightarrow \neg\neg\varphi$
[A4]	$\neg\neg\varphi \rightarrow \varphi$
[A5]	$(\varphi \wedge (\psi \vee \xi)) \rightarrow ((\varphi \wedge \psi) \vee (\varphi \wedge \xi))$
[A6]	$(\varphi \vee (\psi \wedge \neg\psi)) \rightarrow \varphi$
[A7†]	$((\varphi \rightarrow \psi) \wedge (\psi \rightarrow \xi)) \supset (\varphi \rightarrow \xi)$
[A8†]	$(\varphi \rightarrow (\psi \wedge \xi)) \supset (\varphi \rightarrow \psi)$
[A9†]	$((\varphi \rightarrow \xi) \wedge (\psi \rightarrow \zeta)) \supset ((\varphi \wedge \psi) \rightarrow (\xi \wedge \zeta))$
[A10†]	$((\varphi \rightarrow \xi) \wedge (\psi \rightarrow \zeta)) \supset ((\varphi \vee \psi) \rightarrow (\xi \vee \zeta))$
[A11†]	$(\varphi \rightarrow \psi) \supset (\varphi \supset \psi)$
[A12†]	$((\varphi \leftrightarrow \psi) \wedge f(\varphi)) \supset f(\psi)$
[A14†]	$((\neg\varphi \rightarrow \varphi) \wedge (\varphi \rightarrow \psi)) \supset (\neg\psi \rightarrow \psi)$
[D2]	$(\varphi \rightarrow \psi) \supset (\neg\varphi \rightarrow \psi) \rightarrow (\varphi \rightarrow \psi))$
[S1]	$\varphi \rightarrow \mathbf{t}_\varphi$
[S3]	$(\mathbf{t}_\varphi \vee \psi \vee \xi) \rightarrow (\mathbf{t}_\varphi \vee \xi)$

and rules:

[MP1]	$\varphi, \varphi \supset \psi \Rightarrow \psi$
[MP2]	$\varphi, \varphi \rightarrow \psi \Rightarrow \psi$
[ADJ]	$\varphi, \psi \Rightarrow \varphi \wedge \psi$
[MOD]	$\emptyset \Rightarrow \neg\varphi \rightarrow \varphi$ for φ an axiom

Note that this [MOD] rule is essentially a sort of necessitation rule, allowing us to ascribe modal strength to axioms whose main connective is a $\supset$ without ascribing any topic-theoretic properties. Consequence is defined as usual.

3.3 The Canonical Model

The paper (Ferguson, 2023c) used the method of canonical models to prove completeness. Such models will be crucial to characterizing left and right topic sufficient extensions, warranting the review of some salient definitions. An account of the canonical model's accessibility relation follows from:

Definition 6 *For a* S/PAI *theory* Γ, *the set* $\Gamma^\square = \{\varphi \mid \mathbf{t}_\varphi \to \varphi \in \Gamma\}$

This forms the basis of accessibility, leaving us to introduce apparatus to define the canonical model's concept semilattices:

Definition 7 *For a* S/PAI *theory* Γ, $\sim_\Gamma$ *is an equivalence relation such that* $\varphi \sim_\Gamma \psi$ *if* $\mathbf{t}_\varphi \leftrightarrow \mathbf{t}_\psi \in \Gamma$ *and* $[\![\varphi]\!]_\Gamma$ *is the equivalence class induced by* $\sim_\Gamma$.

Definition 8 *For a* S/PAI *theory* Γ, $(\!(\varphi)\!)_\Gamma^\star = \{\Xi \in \Gamma\!\uparrow^\star \mid \varphi \in \Xi\}$ *where* $\Gamma\!\uparrow^\star = \{\Xi \mid \Gamma^\square \subseteq \Xi\}$

Definition 9 *For a* S/PAI *theory* Γ, $[\![\varphi]\!]_\Gamma = \langle [\![\varphi]\!]_\Gamma, (\!(\varphi)\!)_\Gamma^\star \rangle$.

We then introduce the definition of each function $\oplus_\Gamma$ and $-\!\!\circ_\Gamma$ as follows:

Definition 10 *The functions* $\oplus_\Gamma$ *and* $-\!\!\circ_\Gamma$ *are defined so that:*

- $[\![\varphi]\!]_\Gamma \oplus_\Gamma [\![\psi]\!]_\Gamma = [\![\varphi \wedge \psi]\!]_\Gamma$
- $[\![\varphi]\!]_\Gamma -\!\!\circ_\Gamma [\![\psi]\!]_\Gamma = [\![\varphi \to \psi]\!]_\Gamma$

With these functions defined we have introduced all of the definitions necessary to describe the canonical model.

Definition 11 *The* S/PAI *canonical model is* $\mathfrak{M} = \langle W, R, \mathcal{T}, C, \oplus, -\!\!\circ, v, t, h \rangle$ *where*

- $W = \{\Delta \mid \Delta$ *a maximally consistent prime theory*$\}$
- $R = \{\langle \Gamma, \Xi \rangle \mid \Gamma^\square \subseteq \Xi\}$
- $\Gamma \in v(p)$ *iff* $p \in \Gamma$
- *For all* $\Gamma \in W$:

$$\begin{array}{lll}
\mathcal{T}_\Gamma = t_\Gamma[\mathcal{L}] & and & C_\Gamma = c_\Gamma[\mathcal{L}] \\
t_\Gamma(\varphi) = [\![\varphi]\!]_\Gamma & and & c_\Gamma(\varphi) = [\![\varphi]\!]_\Gamma \\
h_{\Gamma,\Xi}([\![\varphi]\!]_\Gamma) = [\![\varphi]\!]_\Xi & and & h_{\Gamma,\Xi}([\![\varphi]\!]_\Gamma) = [\![\varphi]\!]_\Xi
\end{array}$$

Definition 11 allows us to associate each maximally consistent prime S/PAI theory Γ with its own canonical model.

Definition 12 *For* Γ *a maximally consistent prime* S/PAI *theory, its canonical model* $\mathfrak{M}_\Gamma$ *is the submodel of* $\mathfrak{M}$ *where* $W_\Gamma = \{\Delta \mid \Gamma R \Delta\}$.

(Ferguson, 2023c) offers the following two lemmas to be used in Section 4:

Lemma 1 *If* $\Gamma \vdash \varphi \to \psi$ *then* $\Gamma \vdash \mathbf{t}_\varphi \to \mathbf{t}_\psi$

Lemma 2 $[\![\varphi]\!]_\Gamma \leq_\Gamma [\![\psi]\!]_\Gamma$ *iff* $\mathbf{t}_\psi \to \mathbf{t}_\varphi \in \Gamma$

4 Extensions of S/PAI

Among the features of S/PAI, perhaps the most appealing is its *modularity*. The binary $\multimap_w$ of CA/PAI and $\multimap_w$ of S/PAI offer a sufficiently fine degree of control over semantic constraints to represent of a diverse class of topic-theoretic intuitions. The extent of this modularity can be demonstrated by describing extensions of CA/PAI in which several topic-theoretic features received representations in both model-theoretic forms (reflected as conditions on $\multimap$) and proof-theoretic forms (reflected as individual axioms).

The additional sensitivity of the $\multimap$ function introduced for S/PAI licenses the representation of an even wider range of topic-theoretic properties. As before, studying intermediate systems provides an optimal setting in which to investigate the varieties of topic-theoretic constraints that can be accommodated in the S/PAI framework.

4.1 S/PAI as a Subsystem of CA/PAI

We first examine the relationship between state-sensitive S/PAI and conditional-agnostic CA/PAI introduced in (Ferguson, 2023a). Our axiomatization of S/PAI springs from the axiomatic presentation of CA/PAI in (Ferguson, 2023a), modified by adding and dropping particular axioms. Notably, adding to S/PAI the single axiom $[D1]$

$$[D1] \quad (\mathbf{t}_\varphi \leftrightarrow \mathbf{t}_\psi) \supset (\mathbf{t}_{f(\varphi)} \to \mathbf{t}_{f(\psi)})$$

suffices to capture the full logic CA/PAI. Initially, we can provide a proof-theoretic demonstration of this equivalence.

Observation 1 $\quad$ S/PAI $\oplus ((\mathbf{t}_\varphi \leftrightarrow \mathbf{t}_\psi) \supset (\mathbf{t}_{f(\varphi)} \to \mathbf{t}_{f(\psi)})) = $ CA/PAI.

Proof. That S/PAI $\oplus ((\mathbf{t}_\varphi \leftrightarrow \mathbf{t}_\psi) \supset (\mathbf{t}_{f(\varphi)} \to \mathbf{t}_{f(\psi)}))$ is a subsystem of CA/PAI holds by noting that the axioms of the former are all either axioms of the latter or (in the case of $[S1]$ and $[S3]$) provable in CA/PAI.

That the inclusion holds in the other direction follows from noting that every axiom of CA/PAI is an axiom of the extended S/PAI $\oplus ((\mathbf{t}_\varphi \leftrightarrow \mathbf{t}_\psi) \supset (\mathbf{t}_{f(\varphi)} \to \mathbf{t}_{f(\psi)}))$ save for $[A13^\dagger]$. But $[A13^\dagger]$ is used in the completeness proof for CA/PAI only for its use in proving $[S1]$ and $[S3]$, which, as axioms of S/PAI, are provable. $\quad\square$

Additionally, it is worth examining a semantic proof of equivalence, as this will provide a model-theoretic characterization of the distinct assumptions of S/PAI and CA/PAI. To begin, recall that an argument in a function is called *redundant* in case the choice of input for that argument has no bearing on the value of the function. If we wish to translate this feature to the present context, let us introduce some notation for *topic identity*:

Definition 13 *For contents c, d we write $c \sim_t d$ if $\pi_0(c) = \pi_0(d)$*

Then redundancy of the propositional parameters for the arguments for $\multimap$ will have as an analog a property requiring that $\multimap$ cannot distinguish between contents c and d whenever $c \sim_t d$ irrespective of values of $\pi_1(c)$ and $\pi_1(d)$. This allows us to recognize state sufficiency as a type of *state redundancy*.

Definition 14 *An* S/PAI *model is* state sufficient *if for all w:*

$$c \multimap_w e = d \multimap_w f$$

for all $c, d, e, f \in C_w$ where $c \sim_t d$ and $e \sim_t f$.

Then every state sufficient S/PAI model can be associated with a CA/PAI model with the same theory and every CA/PAI can be converted to a state-sensitive model in which all $\multimap_w$ functions are state sufficient.

Observation 2 *The logic* CA/PAI *is characterized by state sufficient* S/PAI *models.*

Proof. Suppose that a CA/PAI model $\mathfrak{C}$ and an S/PAI $\mathfrak{S}$ have identical frames, identical join semilattices for each w, and agree on the initial valuations and topic assignments of propositional atoms. Say that they are *synchronized* in case for each function $\multimap_w^{\mathfrak{C}}$ from $\mathfrak{C}$ and function $\multimap_w^{\mathfrak{S}}$, the following holds for all $a, b \in \mathcal{T}_w$ and $c, d \in C_w$ such that $\pi_0(c) = a$ and $\pi_0(d) = b$:

$$a \multimap_w^{\mathfrak{C}} b = \pi_0(c \multimap_w^{\mathfrak{S}} d)$$

Then it is a simple induction on complexity of formulae to confirm that:

$$\mathfrak{C}, w \Vdash \varphi \text{ iff } \mathfrak{S}, w \Vdash \varphi$$

Of course, every state sufficient S/PAI model has a CA/PAI model with which it is synchronized. Conversely, every CA/PAI model has at least one S/PAI model with which it is synchronized. Consequently, soundness and completeness of an axiom system with respect to CA/PAI models transfers immediately to the class of state sufficient S/PAI models as well. $\square$

This gives a formal confirmation that S/PAI acts as a corrective to the redundancy of states in the determination of topic by the earlier CA/PAI.

4.2 Extensions by Parry's Axioms

As an exercise prior to defining extensions that satisfy left and right topic sufficiency, we now turn to investigating the space of systems intermediate between S/PAI and CA/PAI. This will provide a good exercise in advance of showing how to characterize left and right topic sufficient extensions of S/PAI.

The first illustration will be to examine extensions of S/PAI by the axioms [A7] and [A8] drawn from the full system of Parry's PAI.[4]

[A7] $((\varphi \to \psi) \land (\psi \to \xi)) \to (\varphi \to \xi)$
[A8] $(\varphi \to (\psi \land \xi)) \to (\varphi \to \psi)$

These axioms were chosen due to continuity with (Ferguson, 2023a). The earlier work introduced systems CA/PAI$_7$ and CA/PAI$_8$—extensions of CA/PAI by [A7] and [A8]—and provided characterizations in terms of semantic conditions on conditional agnostic models. It is thus a natural starting place to characterize corresponding S/PAI extensions:

Definition 15   S/PAI$_7$ *and* S/PAI$_8$ *are the closures of* S/PAI *with* [A7] *and* [A8]:

$$\text{S/PAI}_7 = \text{S/PAI} \oplus ((\varphi \to \psi) \land (\psi \to \xi)) \to (\varphi \to \xi)$$
$$\text{S/PAI}_8 = \text{S/PAI} \oplus (\varphi \to (\psi \land \xi)) \to (\varphi \to \psi)$$

Identifying the systems S/PAI$_7$ and S/PAI$_8$ provides an opportunity to examine the degree to which the details of an extension of CA/PAI carry over to the analogous extension of S/PAI. For example, CA/PAI$_7$ and CA/PAI$_8$ are characterized by imposing conditions described as *middle term eliminability* and *right decomposability*, respectively, on the binary $\multimap$. These properties can be generalized to the state-sensitive $\multimap$ to provide characterizations of the respective S/PAI counterparts. The property of middle term eliminability carries over to the present case seemingly without modification:

Definition 16   Call $\multimap_w$ middle term eliminable *if the following condition holds:*

$$d \multimap_w f \leq_w (d \multimap_w e) \oplus_w (e \multimap_w f)$$

for all $d, e, f \in C_w$.

The further property of right decomposability becomes slightly more complicated in the S/PAI setting:

---

[4]*N.b.* the resemblance to [A7†] and [A8†].

Definition 17 *Call* $\multimap_w$ right decomposable *if the following condition holds:*

$$d \multimap_w e \leq_w d \multimap_w (\langle \pi_0(e) \oplus_w \pi_0(f), \pi_1(e) \cap \pi_1(f) \rangle)$$

for all $d, e, f \in C_w$.

Although the present version of right decomposability appears less elegant than middle term eliminability, it remains a relatively natural property, describing the modest requirement that topic inclusion between consequents must be reflected in the topics of the conditionals in which they appear.

We can now proceed to provide classes of models with respect to which these systems are sound and complete, begining with S/PAI_7:

Theorem 1 S/PAI_7 *is characterized by models in which each* $\multimap_w$ *is middle term eliminable.*

Proof. Soundness reduces to showing validity of [A7]; as this is a trivial task, we emphasize completeness. Let Γ be a maximally consistent S/PAI_7 theory. By Lemma 1, $\Gamma \vdash \mathbf{t}_{(\varphi \to \psi) \wedge (\psi \to \xi)} \to \mathbf{t}_{\varphi \to \xi}$ and by Lemma 2, in the canonical model for Γ it holds that $[\![\varphi \to \xi]\!]_\Gamma \leq_\Gamma [\![(\varphi \to \psi) \wedge (\psi \to \xi)]\!]_\Gamma$. Together with definitions, this entails that:

$$[\![\varphi]\!]_\Gamma \multimap_\Gamma [\![\xi]\!]_\Gamma = [\![\varphi \to \xi]\!]_\Gamma \leq_\Gamma [\![(\varphi \to \psi) \wedge (\psi \to \xi)]\!]_\Gamma = ([\![\varphi]\!]_\Gamma \multimap_\Gamma$$
$$[\![\psi]\!]_\Gamma) \oplus_\Gamma ([\![\psi]\!]_\Gamma \multimap_\Gamma [\![\xi]\!]_\Gamma)$$

As $\mathfrak{M}_\Gamma$ is middle term eliminable, such countermodels exist in case $\Gamma \nvdash \zeta$. □

Similar techniques provide a characterization of S/PAI_8:

Theorem 2 S/PAI_8 *is characterized by models in which each* $\multimap_w$ *is right decomposable.*

Proof. Soundness is easily confirmed, so we consider completeness. Let Γ be a maximally consistent S/PAI_8 theory. Then consider three arbitrary contents $[\![\varphi]\!]_\Gamma, [\![\psi]\!]_\Gamma, [\![\Xi]\!]_\Gamma \in C_\Gamma$. By Lemma 1, the hypothesis that $\Gamma \vdash (\varphi \to (\psi \wedge \xi)) \to (\varphi \to \psi)$ entails that $\Gamma \vdash \mathbf{t}_{\varphi \to (\psi \wedge \xi)} \to \mathbf{t}_{\varphi \to \psi}$. By an appeal to Lemma 2, we infer that $[\![\varphi \to \psi]\!]_\Gamma \leq_\Gamma [\![\varphi \to (\psi \wedge \xi)]\!]_\Gamma$, whence:

$$[\![\varphi]\!]_\Gamma \multimap_\Gamma [\![\psi]\!]_\Gamma = [\![\varphi \to \psi]\!]_\Gamma \leq_\Gamma [\![\varphi \to (\psi \wedge \xi)]\!]_\Gamma = [\![\varphi]\!]_\Gamma \multimap_\Gamma [\![\psi \wedge \xi]\!]_\Gamma$$

As $[\![\psi \wedge \xi]\!]_\Gamma = \langle \pi_0([\![\psi]\!]) \oplus_\Gamma \pi_0([\![\xi]\!]), \pi_1([\![\psi]\!]) \cap \pi_1([\![\xi]\!]) \rangle$, by substitution:

$$\llbracket \varphi \rrbracket_\Gamma \multimap_\Gamma \llbracket \psi \rrbracket_\Gamma \leq_\Gamma \llbracket \varphi \rrbracket_\Gamma \multimap_\Gamma \langle \pi_0(\llbracket \psi \rrbracket) \oplus_\Gamma \pi_0(\llbracket \xi \rrbracket), \pi_1(\llbracket \psi \rrbracket) \cap \pi_1(\llbracket \xi \rrbracket) \rangle.$$

Since the points in C_Γ were chosen arbitrarily, this is to say that any S/PAI_8 canonical model is right decomposable. Completeness follows. $\quad\square$

There are a host of additional axioms through which further intermediate systems intermediate can be defined. We will consider one further extension by Parry's axiom [A11]:

[A11] $\quad (\varphi \to \psi) \to (\varphi \supset \psi)$

On its face, [A11] encapsulates a principle establishing *Ampliativity* while proscribing a conditional's *Explicativity*. This is a sort of material principle that parts of topics are never lost through composition. While such a property is doubtful in cases of *e.g.* counterfactual conditionals, there are many scenarios in which such losslessness is quite plausible. Let us define:

Definition 18 $\quad S/PAI_{11} = S/PAI \oplus (\varphi \to \psi) \to (\varphi \supset \psi)$.

The informal ampliativity we had observed in [A11] is given shape with a further semantic property:

Definition 19 *Call a $\multimap_w$ state ampliative if the following condition holds for contents $d, e \in C_w$:*

$$(\pi_0(d) \oplus_w \pi_0(e)) \leq_w (d \multimap_w e)$$

Observation 3 $\quad S/PAI_{11}$ *is characterized by state ampliative models.*

Proof. Let Γ be a maximally consistent S/PAI_{11} theory. Lemmas 1 and 2 once more ensure that the appearance of [A11] in Γ guarantees also the inclusion of $\mathbf{t}_{\varphi\to\psi} \to \mathbf{t}_{\varphi\supset\psi}$. It follows from this that $\llbracket \varphi \supset \psi \rrbracket_\Gamma \leq_\Gamma \llbracket \varphi \to \psi \rrbracket_\Gamma$. Thus, in the canonical model $\mathfrak{M}_\Gamma$,

$$\pi_0(\llbracket \varphi \rrbracket_\Gamma) \oplus_\Gamma \pi_0(\llbracket \psi \rrbracket_\Gamma) = \llbracket \varphi \rrbracket_\Gamma \oplus_\Gamma \llbracket \psi \rrbracket_\Gamma = \llbracket \varphi \supset \psi \rrbracket_\Gamma \leq_\Gamma \llbracket \varphi \to \psi \rrbracket_\Gamma =$$
$$\llbracket \varphi \rrbracket_\Gamma \multimap_\Gamma \llbracket \psi \rrbracket_\Gamma$$

$\multimap_\Gamma$ is therefore state ampliative, whence for any invalid S/PAI_{11} inference a state ampliative countermodel can be given. $\quad\square$

4.3 Left and Right Topic Sufficiency

We are now prepared to examine the extensions corresponding to left and right topic sufficiency. It is worthwhile to pause on an intermediate step. We have defined $\mathsf{S/PAI}$ to be agnostic concerning the degree to which the propositions respectively expressed by an antecedent and consequent constrain the states over which a conditional is evaluated, *i.e.*, that a conditional is *about*.

One might be swayed by Kratzer's remarks in (Kratzer, 1991) identifying the antecedent as the lone "selector" through which the states contributing to a conditional's subject-matter are determined. In such a case—if one is unmoved by Gabbay's examples of (Gabbay, 1972)—a consequent's degenerate contribution to the overall topic will be restricted to its topic alone. This amounts to a restricted version of *Topic Sufficiency* in which a consequent's topic is adequate to determine its contribution (although this might remain insufficient for the antecedent).

Although perhaps a more artificial case, one could take the complementary position by restricting an *antecedent*'s contribution to its topic, leaving the consequent to determine the stative component of the conditional's subject-matter. This might, for example, be a somewhat natural constraint in an abductive setting in which the consequent to be explained serves to pin down the topic, aligning with expectations concerning the intensional conditionals through which the example's abductive activity was communicated.

We can accommodate both positions in the context of $\mathsf{S/PAI}$. Consider the following axioms:

$$[D1_R^+] \quad (\mathbf{t}_\varphi \leftrightarrow \mathbf{t}_\psi) \supset (\mathbf{t}_{\xi \to \varphi} \to \mathbf{t}_{\xi \to \psi})$$
$$[D1_L^+] \quad (\mathbf{t}_\varphi \leftrightarrow \mathbf{t}_\psi) \supset (\mathbf{t}_{\varphi \to \xi} \to \mathbf{t}_{\psi \to \xi})$$

These can be used to define two extensions of $\mathsf{S/PAI}$:

Definition 20 $\mathsf{S/PAI}_R$ and $\mathsf{S/PAI}_L$ *are the closures of* $\mathsf{S/PAI}$ *with* $[D1_R^+]$ *or* $[D1_L^+]$, *respectively,* i.e.:

$$\mathsf{S/PAI}_R = \mathsf{S/PAI} \oplus (\mathbf{t}_\varphi \leftrightarrow \mathbf{t}_\psi) \supset (\mathbf{t}_{\xi \to \varphi} \to \mathbf{t}_{\xi \to \psi})$$
$$\mathsf{S/PAI}_L = \mathsf{S/PAI} \oplus (\mathbf{t}_\varphi \leftrightarrow \mathbf{t}_\psi) \supset (\mathbf{t}_{\varphi \to \xi} \to \mathbf{t}_{\psi \to \xi})$$

The adequacy of these extensions as reflections of the respective positions is made clear upon providing characteristic semantic constraints on $\multimap$.

Definition 21 *A* $\mathsf{S/PAI}$ *model is* right topic adequate *in case for all w and contents $d, e, f \in C_w$ such that $\pi_0(e) = \pi_0(f)$ (i.e. the topic components of contents e and f are identical):*

33

Thomas Macaulay Ferguson

$$d \multimap_w e = d \multimap_w f$$

and is left topic adequate *in case for all such d, e, f:*

$$e \multimap_w d = f \multimap_w d$$

We can see that these restricted notions of *Topic Suffiency* in fact correspond to the respective systems:

Observation 4 *The logic* S/PAI$_R$ *is characterized by right topic adequate* S/PAI *models.*

Proof. Validity of $[D1_R^+]$ is straightforward. If $w \Vdash \mathbf{t}_\varphi \leftrightarrow \mathbf{t}_\psi$, $t_w(\varphi) = t_w(\psi)$ whence $c_w(\varphi)$ and $c_w(\psi)$ will share topic. By right topic adequacy, $c_w(\xi) \multimap_w c_w(\varphi) = c_w(\xi) \multimap_w c_w(\psi)$ for arbitrary ψ, whence $w \Vdash \mathbf{t}_{\xi \to \varphi} \to \mathbf{t}_{\xi \to \psi}$.

For completeness, consider a maximally consistent prime S/PAI$_R$ theory Γ and consider the canonical model $\mathfrak{M}_\Gamma$. Consider φ and ψ such that $[\![\varphi]\!]_\Gamma = [\![\psi]\!]_\Gamma$. This entails that $\mathbf{t}_\varphi \leftrightarrow \mathbf{t}_\psi \in \Gamma$ and by modus ponens on appropriate instances of $[D1_R^+]$, we infer that $\mathbf{t}_{\xi \to \varphi} \leftrightarrow \mathbf{t}_{\xi \to \psi} \in \Gamma$, whence:

$$c_\Gamma(\xi) \multimap_\Gamma c_\Gamma(\varphi) = [\![\xi \to \varphi]\!]_\Gamma = [\![\xi \to \psi]\!]_\Gamma = c_\Gamma(\xi) \multimap_\Gamma c_\Gamma(\psi)$$

As Γ and the contents from C_Γ were chosen arbitrarily, the property holds for $\multimap$ in general, whence any S/PAI$_R$ canonical model is right topic adequate. Completeness follows by standard methods. □

The next observation follows from an identical line of argumentation.

Observation 5 *The logic* S/PAI$_L$ *is characterized by left topic adequate* S/PAI *models.*

These systems of limited topic adequacy were identified as marking an intermediate step on the way to an adequate picture of how the state sensitivity of S/PAI stands with respect to the topic adequacy of CA/PAI.

This intermediacy can be taken in a literal sense in which each of the axioms $[D1_R^+]$ and $[D1_L^+]$ are a half-way mark between S/PAI and CA/PAI. To make this precise, let us formalize the notion of topic sufficiency with the following semantic condition:

Definition 22 *An* S/PAI *model is (simply)* topic adequate *in case for all w and contents $d, e, f, g \in C_w$:*

$$d \multimap_w f = e \multimap_w g$$

34

whenever $\pi_0(d) = \pi_0(e)$ and $\pi_0(f) = \pi_0(g)$.

Definition 23 *Let $\mathfrak{M}$ be a simply topic adequate S/PAI model and let $\mathfrak{N}$ be a CA/PAI model sharing a common frame, topic $\mathcal{T}_w$, functions v, t_w, $\oplus_w$, etc. Identify the respective $\multimap$ functions by $\multimap_w^{\mathfrak{M}}$ and $\multimap_w^{\mathfrak{N}}$. Then $\mathfrak{M}$ and $\mathfrak{N}$ are synched in case for all w, $a, b \in \mathcal{T}_w$, and $c, d \in C_w$:*

$$a \multimap_w^{\mathfrak{N}} b = c \multimap_w^{\mathfrak{M}} d$$

whenever $\pi_0(c) = a$ and $\pi_0(d) = b$.

Lemma 3 *If $\mathfrak{M}$ is a simply topic adequate S/PAI model and $\mathfrak{N}$ is a CA/PAI model and are synched, then for all w:*

$$\mathfrak{M}, w \Vdash \varphi \text{ iff } \mathfrak{N}, w \Vdash \varphi$$

Proof. This follows from a simple recursion on the complexity of φ. □

Observation 6 *The logic CA/PAI is sound and complete with respect to simply topic adequate S/PAI models.*

Proof. Clearly, for every simply topic adequate S/PAI model one can find a CA/PAI model with which it is synched. Conversely, every CA/PAI model is synched with a simply topic adequate S/PAI model. Consequently, by Lemma 3, the soundness and completeness CA/PAI enjoys with respect to its models carries over to simply topic adequate S/PAI models. □

Our analysis of S/PAI_R and S/PAI_L and the properties of right and left topic adequacy now can be consulted to provide a more refined analysis of CA/PAI. First, note that the following relationship holds between simple topic adequacy and left and right topic adequacy.

Lemma 4 *An S/PAI model is simply topic adequate if and only if it is both right and left topic adequate.*

Proof. Left-to-right is immediate. For right-to-left, fix contents in C_w such that $\pi_0(d) = \pi_0(e)$ and $\pi_0(f) = \pi_0(g)$. Then the identity $d \multimap_w f = e \multimap_w f$ follows from left topic adequacy while $e \multimap_w f = e \multimap_w g$ follows from right topic adequacy, whence $d \multimap_w f = e \multimap_w f = e \multimap_w g$. □

CA/PAI can thus be viewed as the consequence of assuming both right and left topic adequacy in tandem, as the following observation records:

Observation 7 CA/PAI *is equivalent to the closure of* S/PAI *with* $[D1_R^\dagger]$ *and* $[D1_L^\dagger]$, i.e.:

$$\mathsf{S/PAI} \oplus (\mathbf{t}_\varphi \leftrightarrow \mathbf{t}_\psi) \supset (\mathbf{t}_{\xi \to \varphi} \to \mathbf{t}_{\xi \to \psi}) \oplus (\mathbf{t}_\varphi \leftrightarrow \mathbf{t}_\psi) \supset (\mathbf{t}_{\varphi \to \xi} \to \mathbf{t}_{\psi \to \xi})$$

5 Concluding Remarks

That the foregoing frames the model in a Parry-like setting is a convenience; the implicit model can be ported to other topic-sensitive contexts. A very natural next step is to combine the foregoing considerations on left and right topic sufficiency with the modifications to topic-sensitive intentional modals in (Ferguson, 2023b) in order to provide more nuanced evaluations of the subject-matter of counterfactuals.

References

Berto, F. (2022). *Topics of Thought*. Oxford: Oxford University Press.

Berto, F., & Hawke, P. (2021). Knowability relative to information. *Mind*, *130*(517), 1–33.

Ferguson, T. M. (2021). The subject-matter of intensional conditionals. In *Proceedings of the Twelfth Smirnov Readings in Logic* (p. 59--63). Moscow: Russian Society of History and Philosophy of Science Publishing House.

Ferguson, T. M. (2023a). Subject-matter and intensional operators I: Conditional-agnostic analytic implication. *Philosophical Studies*, *180*(7), 1849–1879.

Ferguson, T. M. (2023b). Subject-matter and intensional operators II: Applications to the theory of topic-sensitive intentional modals. *Journal of Philosophical Logic*. (To appear)

Ferguson, T. M. (2023c). Subject-matter and intensional operators III: State-sensitive subject-matter and topic sufficiency. *Review of Symbolic Logic*. (To appear)

Fine, K. (1986). Analytic implication. *Notre Dame Journal of Formal Logic*, *27*(2), 169–179.

Gabbay, D. (1972). A general theory of the conditional in terms of a ternary operator. *Theoria*, *38*(3), 97–104.

Kratzer, A. (1991). Conditionals. In A. von Stechow & D. Wunderlich (Eds.), *Semantics: An International Handbook of Contemporary Research* (pp. 651–656). Berlin: Walter de Gruyter.

Mares, E. (2004). *Relevant Logic: A Philosophical Interpretation.* Cambridge: Cambridge University Press.

Parry, W. T. (1968). The logic of C. I. Lewis. In P. A. Schilpp (Ed.), *The Philosophy of C. I. Lewis* (pp. 115–154). La Salle, IL: Open Court.

Thomas Macaulay Ferguson
Rensselaer Polytechnic Institute, Department of Cognitive Science and Czech Academy of Sciences, Institutes of Philosophy and Computer Science
United States and The Czech Republic
E-mail: `tferguson@gradcenter.cuny.edu`

Modality and the Structure of Assertion

ANSTEN KLEV[1]

Abstract: A solid foundation of modal logic requires a clear conception of the notion of modality. Modern modal logic treats modality as a propositional operator. I shall present an alternative according to which modality applies primarily to illocutionary force, that is, to the force, or mood, of a speech act. By a first step of internalization, modality applied at this level is pushed to the level of speech-act content. By a second step of internalization, we reach a propositional operator validating the modal logic S4. After a brief discussion of problematic modality and possibility, the article concludes with an extension of the account that identifies modality with illocutionary force. Throughout, close attention is paid to the intended interpretation of the formalism. All of the rules stipulated will be justified on the basis of this interpretation.

Keywords: foundations of modal logic, speech act theory

1   Introduction

In modern modal logic, modal operators operate on propositions, or on their formal simulacra, well-formed formulas: a modal operator (let us assume that it is unary) takes a proposition, or well-formed formula, and yields another proposition, or well-formed formula. The logico-grammatical category of such an operator is therefore that of a unary propositional connective, another standard example of which is negation. One might well ask whether this treatment of modality is conceptually the most illuminating that logic could offer. English has phrases, such as "that it is necessary", "that it is known", and "that it is possible", that transforms a that-clause into a new that-clause. Taking propositions to be expressed by that-clauses, such phrases could be

---

[1]For comments and discussion, I am grateful to Bruno Bentzen, Will Stafford, Göran Sundholm and Colin Zwanziger. Heinrich Wansing pointed me to Blamey and Humberstone (1991). While writing the paper I have been supported by a *Lumina quaeruntur* fellowship (LQ300092101) from the Czech Academy of Sciences.

taken to express modal operators. There is, however, another conception of modality which, following Kant, we may call the modality of a judgement. The modality of a judgement, as Kant conceived it, does not concern the content of the judgement, but rather the attitude of the judging subject to this content.

In this article I offer an account of modality and the conceptual genealogy of modal logic that may help to clarify the relation between these two conceptions of modality. The account owes much to the so-called judgemental reconstruction of modal logic of Pfenning and Davies (2001) and to some remarks on the syntax of modal logic by Sundholm (2003). In concentrating on the structure of assertion I am following Martin-Löf. The first way in which the account of modality offered here expands on Pfenning and Davies's is in its use of a finer analysis of assertion. (I take it for granted that this analysis is also an analysis of judgement, since I assume judgement and assertion to share the same logical structure. Although the term "assertion" dominates in this article, it can in most places be replaced by "judgement".) The second way is in its offering somewhat different meaning explanations. We are interested in logical systems as meaningful formalisms, to be used for reasoning with, rather than purely mathematical structures that we can only reason about. Meaning in such systems is instituted by certain stipulations that we call meaning explanations. Once basic notions such as assertion and assertoric content have been explained, such stipulations often take the form of meaning-determining rules.

2   The structure of assertion

I shall not attempt a definition of assertion, nor of the more general notion of a speech act. Instead I shall take it for granted that there is a speech act whose primary and designated role is the communication of knowledge, and I shall call that speech act assertion. A clear example is the speech act a mathematician makes when communicating a theorem, be it on the blackboard, in a paper, or over a coffee. An assertion, being an act, takes place in time, yet it has a structure that is not bound to time and that may be the object of theorizing. That at least is a presupposition of speech act theory—and of phenomenology, which studies not only speech acts, but intentionality more broadly. I will follow their lead here.

The fundamental structure of a speech act is the force/content structure. This structure can best be illustrated by series of examples where one element stays constant and the other varies. In normal utterances of the following three sentences,

1. He will leave the room.
2. Will he leave the room?
3. Leave the room!

the content remains the same, but the force varies. That the force varies means simply that the kind of speech act varies, namely from assertion (1) to question (2) to command (3). A series of speech acts of the same kind—with the same force—but with varying content is provided by the series of theorems, major or minor, contained in any mathematical research paper.

A standard assumption of speech act theory is that the content of a speech act is to be identified with a proposition in the sense of modern logic. There are, of course, various ways of understanding the notion of proposition in logic, but on any understanding, it is an object of some kind: a truth value, a set of possible worlds, a type of proof objects, etc. A moment of reflection shows that a proposition understood in any of these ways cannot play the role of the content of a speech act. In particular, it cannot play the role of content of an assertion: one cannot attach assertoric force to a truth value, a set, or a type. What one can attach assertoric force to is the content that a truth value is equal to the truth value True, that the actual world is an element of a set of possible worlds, or that a type of proof objects is non-empty.

Martin-Löf (2003) concluded that we must distinguish the notion of the content of a speech act from the notion of proposition. In the first instance, this gives rise to a three-levelled analysis of assertion:

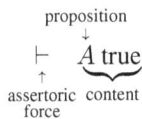

$$
\underset{\substack{\uparrow \\ \text{assertoric} \\ \text{force}}}{\vdash} \overset{\substack{\text{proposition} \\ \downarrow}}{\underset{\text{content}}{\underbrace{A \text{ true}}}}
$$

The word "true" here is not a truth predicate, since "A true" does not stand for a proposition, but for a content. I shall call it the truth particle. It stands for an operation that transforms a proposition into a content. The precise specification of this operation will depend on the underlying notion of proposition. For instance, if A is a truth value, then A true is the content that A is equal to the True. If A is a type of proof objects, as in Martin-Löf's constructive meaning theory, then A true is the content that A is non-empty.

Hare (1970, 1989) observed that Frege's assertion sign—which in Frege's writings is a vertical stroke, but here, in a tradition following the *Principia*

41

Mathematica, is the turnstile—serves a double role: it is both a sign of assertoric force and a sign of what Hare called subscription, that is, a sign to indicate that the assertion has in fact been made. This latter indication is no part of the assertion sign as we shall use it here. For us, the assertion sign will be purely a sign of assertoric force.

The displayed structure is the structure of a categorical assertion. Deductive reasoning consists not only of categorical assertions, but also of hypothetical assertions. A hypothetical assertion is an assertion, hence its force is the assertoric force, but its content has hypothetical form:

$$\vdash \underbrace{A_1 \text{ true}, \ldots, A_n \text{ true} \Rightarrow A \text{ true}}_{\text{content}}$$

The arrow here is not a propositional connective, such as implication, but a content-forming operation, namely an operation that forms hypothetical content. We call contents to the left of the arrow hypotheses and the content to the right the succedent. If the set Γ of hypotheses is empty, then the content $\Gamma \Rightarrow A$ true is just the content A true. If the set of hypotheses is non-empty, the arrow is explained by means of a modus-ponens-like rule, the precise form of which depends on the form of the hypotheses. In the present case, the rule is as follows:

$$\frac{\vdash \Gamma, A \text{ true} \Rightarrow C \qquad \vdash A \text{ true}}{\vdash \Gamma \Rightarrow C}$$

Here and in what follows, C is any content that may serve as succedent. A cut rule can now be justified:

$$\frac{\vdash \Gamma, A \text{ true} \Rightarrow C \qquad \vdash \Gamma' \Rightarrow A \text{ true}}{\vdash \Gamma, \Gamma' \Rightarrow C} \qquad \text{(true-cut)}$$

The justification proceeds by induction on the size of Γ'.

Cut rules such as (true-cut) above and (apod-cut) below are introduced in this article only to facilitate the justification of certain further rules that will be used, in Section 5, in the validation of S4. Whether a cut rule is a primitive rule in the formal system or is merely admissible matters nothing to our purposes. What does matter is that the cut rules introduced are justified by the meaning of the arrow.

We assume that formation rules for propositions are included in the formalism, as they are in Martin-Löf's type theory. There is thus a form of content, A prop, expressing that A is a proposition. (To explain this form of content is just to explain what a proposition is.) We use this form of content in the formulation of two rules that are justified by the meaning-determining rule for the arrow. The first is the rule of assumption:

$$\frac{\vdash A \text{ prop}}{\vdash A \text{ true} \Rightarrow A \text{ true}} \tag{Assm}$$

The second is a rule of weakening:

$$\frac{\vdash \Gamma \Rightarrow A \text{ true} \qquad \vdash B \text{ prop}}{\vdash \Gamma, B \text{ true} \Rightarrow A \text{ true}} \tag{Weak}$$

The justification of the latter proceeds by induction on the size of Γ.

3 Introducing modality

Martin-Löf's three-levelled analysis of assertion gives us three possible operands for modality. I shall outline a view according to which modality applies primarily to the assertion as a whole, or more precisely, to assertoric force. By a process I shall call internalization, following both Pfenning and Davies (2001) and Sundholm (2003), the modality is pushed inwards, first to the level of content, modifying the truth particle, and next to the level of propositions, yielding a modal operator in the usual sense of modal logic.

Letting modality operate primarily on the force of an assertion seems to me to make good sense of Kant's claim that modality contributes nothing to content, but concerns rather the relation between this content and "thought as such" (Kant, 1781/1787, A75/B100). The force of a speech act can naturally be said to correspond to the attitude that the utterer takes to its content—assuming, of course, that the utterance is a sincere one. In the *Ideas* (1913), Husserl dedicated a long discussion (§§ 103–112) to the phenomenon of modified "thetic character", which was Husserl's term in that book for what in effect is a generalization of the notion of force to intentional acts more broadly. Searle and Vanderveken (1985, pp. 63–64) introduce various operations on the forces of speech acts which they designate by means of prefixed boxes, though they do not speak of these operations as modalities.

I shall be concerned, in the first instance, with apodeictic modality. Operating on the force of an assertion, apodeictic modality indicates that the assertion in question has been demonstrated, or proved. (The Greek noun *apodeixis* was used to mean demonstration, both in the sense of showing forth something or someone and in the sense of scientific demonstration.)

I shall not attempt to define the bounds of apodeictic knowledge, but I do take it for granted that mathematical knowledge is, in the typical case, apodeictic. Knowledge supported, not by demonstration, but by testimony, is an instance of what may be called assertoric knowledge. That Caesar crossed the Rubicon I can know only assertorically, since I have no way of going back in time to witness with my own senses that river crossing. The notion of assertoric knowledge turns out to play an essential role in the explanation of the validity of inference (Klev, 202x).

Given this gloss on apodeictic modality, it is natural to postulate the following "necessitation" rule as meaning determining for it:

$$\frac{\begin{array}{c}\mathcal{D}\\ \vdash C\end{array}}{\Vdash C} \qquad (\Vdash\text{-intro})$$

In words this says that the content C, which in general has hypothetical form, $\Gamma \Rightarrow C'$, may be apodeictically asserted provided one has demonstrated the assertion $\vdash C$. A demonstration is a sequence of valid inferences beginning from axioms. The boldface line is used to indicate that this is not a usual form of inference. If a colleague in the mathematics department makes an assertion of the form $\vdash A \wedge B$ true, I may, without further ado, infer $\vdash A$ true and $\vdash B$ true. I may, however, not infer $\Vdash A \wedge B$ true unless I have seen, or otherwise become confident that there is a demonstration of this assertion. The rule of $\Vdash$-introduction is context sensitive in a way rules of inference usually are not. To apply it, one needs to take into account, not only the premiss assertion, but the whole demonstration, $\mathcal{D}$, that precedes it.

4   First internalization

The level of assertion is the appropriate level for the application of apodeictic modality, since what one demonstrates are assertions and not assertoric contents or propositions. We can, however, make good sense of internalizing the apodeictic modality of an assertion by making it part of its content. More precisely, modality at this level can be taken to modify the truth particle. I shall write "apod", short for "apodeictically true", and introduce the form of content A apod. (Pfenning and Davies use "valid" for the same purpose.) The general form of hypothetical content is then

$$A_1 \text{ apod}, \ldots, A_n \text{ apod} \,;\, B_1 \text{ true}, \ldots, B_m \text{ true} \Rightarrow A \text{ true/apod}$$

Writing Δ for the set of apodeictic hypotheses and Γ for the set of truth hypotheses, the same can be written more compactly as follows:

$$\Delta; \Gamma \Rightarrow A \text{ true/apod}$$

The slash is used here to indicate that the form of the succedent is either A true or A apod.

That the apod particle internalizes apodeictically modified force is captured by its introduction rule:

$$\frac{\Vdash A \text{ true}}{\vdash A \text{ apod}} \qquad \text{(apod-intro)}$$

The meaning of the arrow is extended in the obvious way, giving rise to a cut rule and rules of assumption and weakening. The cut rule is as follows:

$$\frac{\vdash \Delta, A \text{ apod} ; \Gamma \Rightarrow C \qquad \vdash \Delta'; \Gamma' \Rightarrow A \text{ apod}}{\vdash \Delta, \Delta'; \Gamma, \Gamma' \Rightarrow C} \qquad \text{(apod-cut)}$$

The peculiar nature of apodeictic hypotheses is captured by the following pair of rules:

$$\frac{\vdash \Delta, A \text{ apod} ; \Gamma \Rightarrow C \qquad \Vdash \Delta; \emptyset \Rightarrow A \text{ true}}{\vdash \Delta; \Gamma \Rightarrow C} \qquad \text{(apod-true-cut)}$$

$$\frac{\Vdash \Delta; \emptyset \Rightarrow A \text{ true}}{\vdash \Delta; \emptyset \Rightarrow A \text{ apod}} \qquad \text{(apod-intro*)}$$

Although (apod-intro*) is in effect a generalization of (apod-intro), it cannot replace the latter as meaning determining for the apod particle, owing to the occurrence of this particle in the hypotheses, Δ.

Justification of (apod-true-cut) and (apod-intro).* We justify these rules simultaneously by induction on the size of the set Δ of apodeictic hypotheses.

For size 0, (apod-intro*) is just (apod-intro), which is valid by stipulation, so consider (apod-true-cut). Its right-hand premiss has the form $\Vdash A$ true, whence we may infer $\vdash A$ apod. Its left-hand premiss has the form $\vdash A$ apod $; \Gamma \Rightarrow C$, whence we may infer $\vdash \Gamma \Rightarrow C$ by (apod-cut).

For size greater than 0, let us first consider (apod-true-cut). We wish to justify inferences of the following form:

$$\frac{\vdash \Delta, B \text{ apod}, A \text{ apod} ; \Gamma \Rightarrow C \qquad \Vdash \Delta, B \text{ apod}; \emptyset \Rightarrow A \text{ true}}{\vdash \Delta, B \text{ apod}; \Gamma \Rightarrow C} \qquad (\gamma)$$

Under the assumption that the two premisses are correct, we must show that the conclusion is correct. We do the latter by showing, firstly, that the following inference is correct:

$$\frac{\vdash B \text{ apod}}{\vdash \Delta; \emptyset \Rightarrow A \text{ apod}} \quad (\delta)$$

Assume, therefore, that $\vdash B$ apod is correct. From the meaning of the apod particle, it follows that also $\Vdash B$ true is correct, whence there is a demonstration of $\vdash B$ true. This demonstration can be extended to a demonstration of $\vdash B$ apod, whence we may infer $\Vdash B$ apod. Using (true-cut) with the usual turnstile, $\vdash$, replaced by the reinforced turnstile, $\Vdash$, we may then infer $\Vdash \Delta; \emptyset \Rightarrow A$ true from the right-hand premiss of (γ). Now we use the induction hypothesis on (apod-intro*) to infer $\vdash \Delta; \emptyset \Rightarrow A$ apod. The justification of (δ) is thus complete. From the meaning of the arrow, it follows that $\vdash \Delta, B$ apod; $\emptyset \Rightarrow A$ apod is correct. From the left-hand premiss of (γ) we may infer $\vdash \Delta, B$ apod; $\Gamma \Rightarrow C$ by (apod-cut). This completes the induction step for (apod-true-cut).

The induction step for (apod-intro*) follows immediately by letting Γ be $\emptyset$ and C be A apod. $\quad\square$

One more rule pertaining to the apod particle is needed for the development of S4 in the following section. From the meaning-determining role of (apod-intro) it is clear that the following rule is justified:

$$\frac{\vdash \Delta; \Gamma \Rightarrow A \text{ apod}}{\vdash \Delta; \Gamma \Rightarrow A \text{ true}} \quad \text{(apod-elim)}$$

Rules governing the propositional connectives are formulated only by means of the truth particle, not by means of the apod article. For instance, the rules for implication are as follows:

$$\frac{\vdash \Gamma, A \text{ true} \Rightarrow B \text{ true}}{\vdash \Gamma \Rightarrow A \supset B \text{ true}} \qquad \frac{\vdash \Gamma \Rightarrow A \supset B \text{ true} \quad \vdash \Gamma' \Rightarrow A \text{ true}}{\vdash \Gamma, \Gamma' \Rightarrow B \text{ true}}$$

As will be seen in an example in the next section, one can use (apod-elim) and (apod-intro) to analyze propositions under the apod particle.

In this section and the next we have relied on Pfenning and Davies (2001, § 4), though with some deviations. Three deviations are worth noticing. The three-levelled analysis of assertion led us to introduce the force operator $\Vdash$, not used by Pfenning and Davies. Unlike them, we have allowed the apod particle to occur in the succedent, not only in the hypotheses. Finally, we have offered a justification of, and not simply stipulated, the rule (apod-true-cut).

5 Second internalization

The second internalization introduces modality as a propositional operator. We stipulate, first of all, that $\Box A$ is a proposition whenever A is a proposition. That $\Box$ is indeed the internalization of the apod particle is clear from its meaning-determining introduction rule:

$$\frac{\vdash A \text{ apod}}{\vdash \Box A \text{ true}} \qquad (\Box\text{-intro})$$

From the meaning of the arrow, the corresponding rule that includes hypotheses is justified:

$$\frac{\vdash \Delta;\Gamma \Rightarrow A \text{ apod}}{\vdash \Delta;\Gamma \Rightarrow \Box A}$$

No restrictions are placed on the set Δ,Γ of hypotheses here, and it remains unchanged in the passage from premiss to conclusion. This is uncommon in natural-deduction formulations of modal logic (cf. e.g. Prawitz, 1965, ch. vi). A typical and easily formulable restriction says—translated to the present framework—that all hypotheses must have the form $\Box B$ true. Such a restriction works well formally, but places the meaning-determining role of ($\Box$-intro) in jeopardy, owing to the negative occurrences of $\Box$ in its premiss.

For the elimination rule we may use a natural deduction formulation:

$$\frac{\Box A \text{ true} \qquad \begin{array}{c} (A \text{ apod}) \\ \mathcal{D} \\ C \end{array}}{C} \qquad (\Box\text{-elim})$$

This elimination rule follows a pattern familiar, for instance, from $\vee$-elimination. It is justified through the stipulation of the following reduction:

$$\frac{\dfrac{\begin{array}{c}\mathcal{D}\\ A \text{ apod}\end{array}}{\Box A \text{ true}} \qquad \begin{array}{c}(A \text{ apod})\\ \mathcal{D}'\\ C\end{array}}{C} \quad\rightsquigarrow\quad \begin{array}{c}\mathcal{D}\\ \mathcal{D}'\\ C\end{array} \qquad (\Box\text{-red})$$

The right-hand derivation here is obtained by replacing each occurrence of A apod in leaf position in $\mathcal{D}'$ by the derivation $\mathcal{D}$, whose conclusion is A apod. Such a replacement is justified by the rule (apod-cut).

47

For illustration we display, in a natural deduction format, a demonstration that the K scheme is true:

$$
\cfrac{
  \Box(A \supset B) \text{ true}^1 \qquad
  \cfrac{
    \Box A \text{ true}^2 \qquad
    \cfrac{
      \cfrac{
        \cfrac{A \supset B \text{ apod}^3}{A \supset B \text{ true}} \qquad
        \cfrac{A \text{ apod}^4}{A \text{ true}}
      }{
        \cfrac{\cfrac{B \text{ true}}{B \text{ apod}}}{\Box B \text{ true}} \; 4
      }
    }{
      \cfrac{\Box B \text{ true}}{\Box A \supset \Box B \text{ true}} \; 2
    }
  }{\Box A \supset \Box B \text{ true}} \; 3
}{\Box(A \supset B) \supset (\Box A \supset \Box B) \text{ true}} \; 1
$$

The dotted line indicates two inference steps which we can make explicit by writing out the first part of this derivation as a proper demonstration:

$$
\cfrac{
  \cfrac{\vdash A \supset B \text{ apod} \Rightarrow A \supset B \text{ apod}}{\vdash A \supset B \text{ apod} \Rightarrow A \supset B \text{ true}} \qquad
  \cfrac{\vdash A \text{ apod} \Rightarrow A \text{ apod}}{\vdash A \text{ apod} \Rightarrow A \text{ true}}
}{
  \cfrac{
    \cfrac{\vdash A \supset B \text{ apod}, A \text{ apod} \Rightarrow B \text{ true}}{\Vdash A \supset B \text{ apod}, A \text{ apod} \Rightarrow B \text{ true}}
  }{\vdash A \supset B \text{ apod}, A \text{ apod} \Rightarrow B \text{ apod}}
}
$$

A boldface inference line is used to indicate application of ($\Vdash$-intro) from the third to the fourth line. As already noted, application of this rule requires that we reflect on the whole of the preceding demonstration, not only the premiss.

The derivations of $\vdash \Box A \supset A$ true and $\vdash \Box A \supset \Box\Box A$ true employ the same ideas. Thus, all the defining schemes of S4 are derivable.

Our formulation of ($\Box$-intro) relies on the possibility of using the apod particle in the succedent. In Pfenning and Davies's formulation, the premiss of ($\Box$-intro) is rather $\Delta; \cdot \Rightarrow A$ true, rendering their $\Box$-introduction similar to the rule (apod-intro*) from the previous section. The present formulation ($\Box$-intro) has been preferred here for two reasons: it explicitly introduces $\Box$ as an internalization of apod, and it fits perfectly with ($\Box$-elim) as elimination rule.

The idea of distinguishing hypotheses and succedents of two kinds in sequent calculi for modal logic was explored already by Blamey and Humberstone (1991). They did not introduce any novel logico-grammatical categories—such as our contents and assertions—but operated just with formulae, corresponding to our propositions. They did, however, remark (p. 776) that "the move from truth-functional to modal logic" may best be made, not "by adding a new primitive connective with new rules governing it, but rather by extending one's conception of the objects to be manipulated by such rules." In the present terminology, the objects manipulated by such rules are indeed contents rather than propositions.

6   Problematic modality

Besides apodeictic and assertoric modality, the third and last modality of judgement recognized by Kant is the problematic modality. In Kant's table of categories, which he took to be derived from his table of judgements, problematic modality corresponds to possibility. From Kant's discussion of problematic modality, it appears that the conception of possibility here must be a rather weak one. An especially instructive passage is the following (A75/B101):

> The problematic proposition [*Satz*] is therefore that which only expresses logical possibility (which is not objective), i.e., a free choice to allow such a proposition to count as valid, a merely arbitrary assumption [*Aufnehmung*] of it in the understanding.

Transferred to the current setting, this passage (and others from the same section of the *Critique*) suggests to me that *any* proposition in the present sense may be judged with problematic modality to be true. Every proposition in the present sense "expresses logical possibility" in the sense that it may be assumed to be true.

It is therefore natural to stipulate the following rule as meaning determining for problematic modality:

$$\frac{\vdash A \text{ prop}}{\wr A \text{ true}} \qquad (\wr\text{-intro})$$

Problematically modified assertoric force is thus indicated by means of a squiggly turnstile. Its applicability is restricted here to content of the form A true. Restriction on the form of content to which a given force can apply is a familiar phenomenon in speech act theory (cf. the so-called propositional content rule of Searle, 1969, p. 63.) I shall not discuss whether introduction rules for problematic modality can be given that is less restrictive as to the form of the operand. Meaning-determining rules for problematic modality are not without interest, since $\wr C$ in effect asserts that C is a content.

We capture the tight relation that Kant appears to have seen between problematic modality and the notion of assumption by laying down the following elimination rule:

$$\frac{\vdash A \text{ true}}{\vdash A \text{ true} \Rightarrow A \text{ true}} \qquad (\vdash\text{-elim})$$

It will be clear that a force regulated by these two rules is of little use: an assertion of the form $\vdash A$ prop already provides for what we might want to do with the problematically modified assertion $\vdash A$ true. It is sometimes asked whether there is a separate speech act of assumption. We might call $\vdash$ the force of such a speech act and gloss $\vdash A$ true as entertaining A to be true. The logical use of assumptions, however, comes out properly only through the occurrence of contents as hypotheses in assertions of hypothetical form.

The first internalization of $\vdash$ gives us just the form of content A prop, and the second internalization gives an operator π with the introduction rule

$$\frac{\vdash A \text{ prop}}{\vdash \pi A \text{ true}}$$

Such an operator is of little interest here, but it may be of interest in a formal language that allows for the formation of sentences deemed meaningless.

7 Possibility

Problematic modality, in particular its second internalization, π, is quite far removed from what standard modal logic would make one expect of a possibility modality. In particular, one would not expect in standard modal logic to find $\lozenge A$ to be true for every proposition A. Pfenning and Davies (2001, § 5) extended their account of modal logic to a more standard possibility modality by relying on possible-worlds intuitions. The account follows the same pattern as their treatment of necessity: it begins with rules for a content-forming particle, poss, which is then employed for the formulation of natural deduction rules.

The particle poss operates on a proposition A to form the content A poss. Where Γ and Δ are as before, and E is a—possibly empty—set of hypotheses of the form B poss, the general form of hypothetical content is now

$$E; \Delta; \Gamma \Rightarrow A \text{ true/apod/poss}$$

The cut rule for poss is the obvious one, and rules of assumption and weakening follow.

Pfenning and Davies lay down the following two rules for the poss particle:

$$\frac{\vdash A \text{ true}}{\vdash A \text{ poss}} \qquad \frac{\vdash E; \Delta; \Gamma \Rightarrow A \text{ poss} \qquad \vdash \Delta; A \text{ true} \Rightarrow B \text{ poss}}{\vdash E; \Delta; \Gamma \Rightarrow B \text{ poss}}$$

It is natural to regard both of these as (meaning-determining) introduction rules. The first rule gives the base case for the application of poss, and the second rule, the successor case. In the second premiss of the second rule, it is essential that A true is the only truth hypotheses and that there are no possibility hypotheses. If A poss is taken to mean that there is a world in which A is true, then the second rule may be compared to $\exists$-elimination, and the hypothesis A true may be compared to the hypothesis that a world w is given where A is true. To be allowed to use this hypothesis in an application of $\exists$-application, we cannot assume to know anything more about w than this, whence the restriction on the hypotheses in the second premiss. From the meaning of the arrow it follows that the first rule remains justified under the extension of the content by arbitrary hypotheses.

Having introduced the poss particle, we can use it to give natural deduction rules for $\Diamond$ rendering it the internalization of poss. The rules are precise parallels to the rules for $\Box$:

$$\frac{\vdash A \text{ poss}}{\vdash \Diamond A \text{ true}} \ \Diamond\text{-intro} \qquad\qquad \frac{\Diamond A \text{ true} \qquad \begin{array}{c} (A \text{ poss}) \\ \mathcal{D} \\ C \end{array}}{C} \ \Diamond\text{-elim}$$

The corresponding reduction rule is justified by the cut rule for the poss particle. Using these rules, one can derive $\vdash A \supset \Diamond A$ true, $\vdash \Diamond\Diamond A \supset \Diamond A$ true, and $\vdash \Box(A \supset B) \supset (\Diamond A \supset \Diamond B)$ true.

Whereas the diamond, $\Diamond$, is thus rendered the internalization of the poss particle, it is difficult to see from the two introduction rules for poss that it could be rendered the internalization of a modification of assertoric force or indeed of any illocutionary force. The second premiss of the second rule poses a problem for such a rendering, since poss there occurs in the succedent, but is not allowed to occur in the hypotheses. Since the introduction rules for poss relied on possible-worlds intuitions rather than speech act theory, it is perhaps only to be expected that no such rendering is forthcoming.

8 Generalizing the account

Let Φ be an illocutionary force that is applicable to content of the form A true, perhaps with some restriction on the proposition A. A natural question is then whether, following our account of apodeictic and problematic modality, two steps of internalization can be carried out:

$$\Phi \; A \text{ true} \quad \longrightarrow \quad \vdash A \text{ phi} \quad \longrightarrow \quad \vdash \phi A \text{ true}$$

Not every modal operator of standard modal logic can be regarded as such a ϕ. The operators of temporal logic are a clear example. The possibility modality described in the previous section, motivated by possible-worlds intuitions, may be another. These modal operators are, however, not counterexamples to the thesis that every force Φ of the kind described gives rise to a modal operator ϕ.

One might ask whether plain assertoric force, $\vdash$, is such a counterexample. It is not. There is a clear sense in which the truth particle plays a role at the level of content similar to that played by assertoric force at the level of force: each of them is the standard form with respect to modalization. Under the first internalization, modified assertoric force is replaced by unmodified assertoric force, $\vdash$, and under the the second internalization, a modified truth particle is replaced by the unmodified truth particle, true. The truth particle is thus naturally seen as the internalization of assertoric force. The internalization of the truth particle, in turn, is the truth modality, T:

$$\frac{\vdash A \text{ true}}{\vdash T(A) \text{ true}} \qquad \qquad (T\text{-intro})$$

Its elimination rule may be formulated simply as the inverse of this introduction rule:

$$\frac{\vdash T(A) \text{ true}}{\vdash A \text{ true}} \qquad \qquad (T\text{-elim})$$

As one would expect of the internalization of the truth particle, the truth modality is neutral and leaves everything as it was. The second internalization, quite generally, offers a way of expressing a modified truth particle in terms of the unmodified truth particle, but with a modified proposition. It is clear that, in order to express A true in terms of the truth particle, no changes are needed to the proposition A.

The view suggested is thus that every illocutionary force applicable to content of the form A true gives rise to a modal operator. Such a view in turn suggests that there are deep connections between modal logic and speech act theory. Investigating those connections further could be beneficial to the philosophical study of modal logic.

References

Blamey, S., & Humberstone, L. (1991). A perspective on modal sequent logic. *Publications of the Research Institute for Mathematical Sciences, Kyoto University*, *27*, 763–782.

Hare, R. M. (1970). Meaning and speech acts. *The Philosophical Review*, *79*, 3–24.

Hare, R. M. (1989). Some sub-atomic particles of logic. *Mind*, *98*, 23–37.

Husserl, E. (1913). *Ideen zu einer reinen Phänomenologie und phänomenologischen Philosophie. Erstes Buch*. Halle: Max Niemeyer.

Kant, I. (1781/1787). *Kritik der reinen Vernunft*. Riga: Johann Friedrich Hartknoch. (Quoted from the English translation of Guyer and Wood (Cambridge, 1998))

Klev, A. (202x). Martin-Löf on the validity of inference. In A. Piccolomini d'Aragona (Ed.), *Perspectives on Deduction*. Cham: Springer. (Forthcoming)

Martin-Löf, P. (2003). Are the objects of propositional attitudes propositions in the sense of propositional and predicate logic? (Transcript of a lecture given in Geneva on 10 December 2003. Available at https://pml.flu.cas.cz/)

Pfenning, F., & Davies, R. (2001). A judgemental reconstruction of modal logic. *Mathematical Structures in Computer Science*, *11*, 511–540.

Prawitz, D. (1965). *Natural Deduction*. Stockholm: Almqvist & Wiksell.

Searle, J. R. (1969). *Speech Acts*. Cambridge: Cambridge University Press.

Searle, J. R., & Vanderveken, D. (1985). *Foundations of Illocutionary Logic*. Cambridge: Cambridge University Press.

Sundholm, B. G. (2003). "Mind your P's and Q's!" On the proper interpretation of modal logic. In T. Childers & O. Majer (Eds.), *LOGICA Yearbook 2002* (p. 233–243). Prague: Filosofia.

Ansten Klev
Czech Academy of Sciences, Institute of Philosophy
The Czech Republic
E-mail: klev@flu.cas.cz

Abelard and the Development of Connexive Logic

WOLFGANG LENZEN[1]

Abstract: In this paper it is argued that Abelard was probably the first logician who tried to defend "hardcore connexivism". He wanted to avoid the "paradoxes" of strict implication by requiring that the consequent must be "contained" in the antecedent. In particular he denied that a negative property like not being a stone is ever contained in a positive property like being a man. However, Alberic of Paris invented a counterexample to show that, on the basis of logical principles accepted by Abelard, the impossible antecedent 'Socrates is a man and Socrates is not an animal' entails its own negation. The paper examines several ways out to escape "Alberic's trap" as they have been attempted by medieval and by modern logicians like E. Nelson.

Keywords: Abelard, Alberic of Paris, E. Nelson, Ex contradictorio quodlibet, connexive logic, relevant logic

1 Introduction

According to (King & Arlig, 2018), Abelard "was the greatest logician since Antiquity: he [...] worked out a complete theory of entailment [...] (which we now take as the theory of logical consequence)".[2] Yet Abelard can hardly be considered as the founder of *connexive logic*.[3] This attribute is usually granted to Aristotle and to Boethius. Thus, with reference to a famous passage from *Prior Analytics*, McCall explains:

> What Aristotle is trying to show here is that two implications of the form 'If p then q' and 'If not-p then q' cannot both be true.

---

[1]I would like to thank two anonymous referees of a former version of this paper for their helpful suggestions.

[2]Martin (2004), p. 158 and p. 166 similarly maintained: "Abelard was the greatest logician between Aristotle and the Stoics in antiquity and William of Ockham and John Buridan in the fourteenth century". "[H]e must be recognized as one of the greatest of all philosophical logicians". For a comprehensive reconstruction of Abelard's logic cf. (Lenzen, 2021).

[3]For a definition of connexive logic cf. (Wansing, 2020).

The first yields, by contraposition, 'If not-q then not-p', and this
together with the second gives 'If not-q then q' by transitivity.
But, Aristotle says, this is impossible: a proposition cannot be
implied by its own negation. [...] We shall henceforth refer
to the principle that no proposition can be implied by its own
negation [...] as *Aristotle's [first] thesis* [...]. The [other]
connexive principle $\neg((p \rightarrow q) \wedge (\neg p \rightarrow q))$ will be referred to
as *Aristotle's second thesis*.

Nine centuries later, we find in Boethius' *De hypotheticis syl-
logismis* [...] an elaborate system of inference-schemata [...]
among them [...] an analogue of Aristotle's second thesis. [It
...] may be transliterated thus:

> If p, then if q, then r,
>
> if q then not-r,
>
> therefore, not-p.

The reasoning that led Boethius to assert the validity of this
schema was presumably this. Since the two implications 'If q
then r' and 'If q then not-r' are incompatible, the second premiss contradicts the consequent of the first premiss. Hence, by
modus tollens, we get the negation of the antecedent of the first
premise, namely 'not-p'. [...] The corresponding conditional,
'If $q \rightarrow r$ then $\neg(q \rightarrow \neg r)$' will be denoted *Boethius' thesis*, and
serves with the thesis $\neg(p \rightarrow \neg p)$ as the distinguishing mark of
connexive logic.[4]

Despite these undeniable facts, Abelard played an important role for the
development of connexive logic. As has been argued in (Lenzen, 2020a),
Aristotle and Boethius most likely understood their connexive theses only in
the sense of a "*humble* connexivism", but Abelard appears to have been the
first logician to defend "*hardcore* connexivism".[5] To wit, humble connexivism can be characterized by the (entirely plausible) conditions

---

[4](McCall, 2012), pp. 415–416. McCall's symbols for negation and conjunction '$\sim$' and
'&' have been replaced by '$\neg$' and '$\wedge$', respectively. Furthermore, we use '$\vee$' as a symbol for
disjunction and '$\rightarrow$' for implication. Since there are different interpretations of implication,
they will be distinguished, if necessary, by subscripts like '$\rightarrow_{str}$' for strict implication, etc.

[5]This term 'humble connexivism' was coined in (Kapsner, 2019), while the term 'hardcore
connexivity' was introduced in (Lenzen, 2019). Let it be emphasized that this term was never
meant to have a degrading or insulting connotation.

- that a *self-consistent* antecedent p cannot imply *its own negation*,

- that a *self-consistent* p cannot imply *both of two contradictory propositions*,

- that a *self-consistent* p cannot therefore imply *every proposition*.

Hardcore connexivism, in contrast, transcends these requirements by maintaining that also

- a *self-inconsistent* proposition doesn't imply every (other) proposition

- in particular, the *self-inconsistent conjunction* $(p \wedge \neg p)$ doesn't imply the single conjuncts p and $\neg p$;

- $(p \wedge \neg p)$ does not even imply its own (tautological) negation $\neg (p \wedge \neg p)$!

One aim of this paper is to investigate Abelard's attempt to defend hardcore connexivism. The basic idea of this defence, and of its eventual failure, was summarized in (King & Arlig, 2018) as follows:

> The key [...] is found in *inferentia*, best rendered as 'entailment' since Abelard requires the connection between the propositions involved to be both necessary and relevant. [...] Abelard spends a great deal of effort to explore the complexities of the theory of topical inference. [...] In the end, it seems that Abelard's principles of topical inference do not work, a fact that became evident with regard to the topic 'from opposites': Abelard's principles lead to inconsistent results.

This issue will be discussed in Sections 2–5 below. The remaining sections are devoted to several medieval (and contemporary) attempts to escape "Alberic's trap". Already in the 12th century, Alberic of Paris discovered that hardcore connexivism is incompatible with the standard laws of conjunction and disjunction.

2 Abelard and the basic principles of connexive implication

In accordance with McCall's explanation (quoted above), *Aristotle's* theses shall be denoted and formalized as:

ARIST1 $\quad \neg(\neg p \to p)$

ARIST2 $\quad \neg((p \to q) \land (\neg p \to q))$

Following Martin, however, the subsequent variants shall not be referred to as *Boethius'* theses, but rather as *Abelard's* theses:

ABEL1 $\quad \neg(p \to \neg p)$

ABEL2 $\quad \neg((p \to q) \land (p \to \neg q)).$[6]

An anonymous referee pointed out that there exist different ways to *formalize* the idea behind principle ABEL2. Perhaps the problems to be discussed in what follows might be avoided by choosing instead of the (negative) *conjunction* ABEL2 the form of an implication like $(p \to q) \to \neg(p \to \neg q)$. As a matter of fact, e.g., Pizzi and Williamson (1997) investigated the latter formula which they refer to as "Strong Boethius' Thesis", while Pizzi (1977) primarily dealt with formula ABEL2, or the equivalent material implication $(p \to q) \supset \neg(p \to \neg q)$. Now, no matter whether the main operator of the formula $(p \to q) \to \neg(p \to \neg q)$ is interpreted as a strict implication or as a "consequential" implication à la Pizzi, it entails at any rate the material variant ABEL2. As will be shown below, in conjunction with some further logical principles which Abelard considered as indispensable, already the weak variant ABEL2 suffices to derive inconsistencies.

Abelard defended ABEL1 (and ARIST1) by saying that "the truth of one of two contradictories doesn't require the truth of the other; instead, it expels and destroys it".[7] Furthermore, he defended ABEL2 (together with ARIST2) by maintaining that "just as the affirmation and the negation of one and the same proposition cannot follow from the same proposition, so also [...] testifying Aristotle's theses, [...] the same consequent cannot follow from the affirmation and from the negation of the same antecedent".[8] Abelard illustrated these principles by arguing that, e.g., 'If something is a man, it's

---

[6]Cf. (Martin, 1987), p. 389: "Abaelard's own principles are: /ABAELARD'S FIRST PRINCIPLE (AB1)/ From no statement can follow both another statement and also the negation of that statement. [...] ABAELARD'S SECOND PRINCIPLE (AB2)/ From no statement can follow its own negation". Note that in our terminology Martin's AB1 is ABEL2 while his AB2 is our ABEL1!

[7]Cf. (Abaelardus, 1970), p. 290: "[...] cum alterius dividentium veritas non solum veritatem alterius non exigat, immo omnino eam expellat et extinguat".

[8]Cf. (Abaelardus, 1970), p. 290: "Neque enim potest esse ut ad idem consequantur affirmatio et negatio de eodem, sicut [...] Aristoteli testante [...] ad affirmationem et negationem eiusdem non sequitur idem consequens".

an animal' and 'If it's not a man, it's an animal' cannot *both* be true because otherwise one might derive the "inconveniency" 'If x is not an animal, x is an animal' (*"si non est animal, est animal"*). Similarly, 'If x is a man, x is an animal' and 'If x is a man, x is not an animal' cannot be true together because otherwise it would follow 'If x is a man, x is not a man'.

3   Strict vs. relevant implication

In part III/1 of *Dialectica*, Abelard distinguishes two kinds of the *necessity of a consequence* ("necessitas consecutionis"). In a more liberal sense, a conditional is true if "that, what is said by the antecedent, cannot be the case without that which is said by the consequent".[9] This requirement corresponds to the modern conception of a strict implication according to which 'If p, then q' is true if and only if ('iff', for short) it is impossible that the antecedent p be true and yet the consequent q be false:

STRICT $(p \rightarrow_{str} q)$ iff $\neg\Diamond(p \wedge \neg q)$.

However, this definition entails certain "inconvenientia" which are nowadays referred to as 'paradoxes of strict implication'. Abelard clearly recognized that if an antecedent is *impossible*, then it is a fortiori impossible that the antecedent be true and yet the consequent be false. Hence one obtains a principle which medieval logicians came to call 'Ex impossibili [sequitur] quodlibet':

EIQ   If $\neg\Diamond p$, then $(p \rightarrow_{str} q)$ for any arbitrary q.

Abelard considered in particular the following instance:

(i)   If Socrates is a stone, then Socrates is an ass.

Both the antecedent and the consequent of (i) are "impossible"; but the impossibility of the antecedent *alone* suffices to conclude, on the basis of STRICT, that (i) is *true*.[10] Abelard, however, regarded (i) as false. The falsity of (i), though, is not founded in the *impossibility* of its components. E.g., Abelard was willing to accept the truth of conditionals like:

---

[9]Cf. (Abaelardus, 1970), p. 283: "Videntur autem due consecutionis necessitates: una quidem largior, cum videlicet id quod dicit antecedens non potest esse absque eo quod dicit consequens [...]".

[10]As Kneale and Kneale (1962, p. 217) remarked: "For it is impossible that Socrates should be a stone, and so impossible that he should be a stone without being an ass".

(ii)   If Socrates is an ass, then Socrates is an animal

(iii)   If Socrates is a pearl, then Socrates is a stone.[11]

While the properties expressed in the consequents of (ii) and of (iii) are "natural" implications of the properties expressed in the respective antecedent, this does not hold in the case of (i). Clearly, being an *ass* is not a "natural" implication of being a *stone*!

More generally, Abelard suggested a stronger requirement for the truth of a conditional, according to which "not only the antecedent cannot be true without the consequent, but also the [truth of the] antecedent requires the [truth of the] consequent *by itself*".[12] Martin (2004), p. 181, paraphrased this condition as follows:

> The antecedent is required to be *relevant* to the consequent in that its truth is genuinely sufficient for that of the consequent and this is guaranteed by the consequent being in some way contained in the antecedent [my emphasis].

According to the "*locus a genere*", Abelard's more demanding conception of a "natural" or relevant implication holds, in particular, when the antecedent contains the concept of a *species* and the consequent contains that of the corresponding *kind*, e.g., in the affirmative case:

(iv)   If x is a man, x is an animal;[13]

or, in the negative case,

(v)   "If this stone is not an animal, it is not a man".[14]

---

[11]Example (ii) is given in (Abaelardus, 1970), p. 346: "Licet enim hec consequentia: '*Si Socrates est asinus, est animal*' antecedens habeat impossibile, tamen non ideo minus est vera". Example (iii) is stated in (Geyer, 1919–1933), p. 329: "Si Socrates est margarita, Socrates est lapis".

[12]Cf. (Abaelardus, 1970), p. 284: "[...] altero vero strictior, cum scilicet non solum antecedens absque consequenti non potest esse verum, <sed etiam> ex se ipsum exigit". Kneale and Kneale (1962) interpreted this condition so that "the antecedent of a true conditional statement requires the consequent *intrinsically*" (p. 217; my emphasis).

[13]Cf. (Abaelardus, 1970), p. 284: "que quidem necessitas in propria consecutionis sententia consistit et veritatem tenet incommutabilem ut, cum dicitur:/ '*si est homo, est animal*'".

[14]Cf. (Abaelardus, 1970), p. 340: "*De loco a genere*/ Tres autem regulas a genere in usum duximus, has quidem/ a quocumque removetur genus, et species [...] 'si hic lapis non est animal, non est homo'".

4 Abelard's rejection of the "locus ab oppositis"

The "topic of opposites" comprises several related principles which Abelard discusses at great length in part III/1 of *Dialectica*.[15] The most pertinent principle deals with opposite *concepts* such as 'man' and 'horse'. The traditional rule governing such concepts says: "If one of the opposites is affirmed, the other has to be denied".[16] Just as the contrariety of two *propositions* p, q, is usually defined by the condition that is it *impossible* that p and q are *together true* (while it may well be that both are false), so the contrariety of *concepts* might be defined as follows:

> Terms A and B are contrary iff it is impossible that there exists an x such that x is A and x is B (while it is possible that there exists an x such that x neither is A nor is B).

Given this definition, it would appear reasonable to suppose that since 'man' and 'stone' are contrary concepts, the corresponding propositions 'Socrates is a man' and 'Socrates is a stone' must be contrary, too. In particular, the following conditional may plausibly be considered as true:

> (vi)   If Socrates is a man, Socrates is not a stone.

However, Abelard wasn't willing to accept this! Martin (1987, p. 392) tried to explain and justify Abelard's position as follows:

> Not being a stone does not follow in the appropriate way from being a man, even though it is inseparable from being a man. It does not follow in the appropriate way since it is no part of the nature of a man that he not be a stone.

In general, although the antecedent of a proposition like (vi) *strictly implies* the consequent, the implication fails to be "natural" because a *negative* concept like 'not-stone' doesn't denote a *natural kind* from which a species like 'man' might be singled out by a corresponding "differentia specifica".[17] More generally, Abelard appears to have believed "that no conditional of

---

[15]Cf. (Abaelardus, 1970), sections "De oppositis" (pp. 369-70); "De contrariis" (pp. 374-384), "De affirmatione et negatione" (pp. 389-390) and "De loco ab oppositis" (pp. 393-397).

[16]Cf. (Abaelardus, 1970), p. 394: "(posito uno oppositorum tollitur alterum)". Or, as Martin (1987, p. 391) puts it, "Opposites have the property that they cannot simultaneously inhere in a substance. If one is present in them the other must be absent."

[17]As Martin (2004, p. 191) puts it: "Since, according to Abelard, there are no negative substantial forms, the definition of a natural kind cannot contain a negative term".

the form '$p \to \neg q$' can be true"[18], and Martin went so far as to argue that Abelard *therefore* rejected the law of double negation, $p \to \neg\neg p$.[19] Anyway, the main reason for Abelard's rejection of (vi) derives from his wish to cope with certain "inconvenientia" that had been discovered by clever logicians of his time.

5   Two "embarrassing arguments"

In his pioneering papers (1986), (1987), Martin reconstructed the following argument:

1. If Socrates is a man and a stone, Socrates is a man.

2. If Socrates is a man, Socrates is not a stone.

So 3. If Socrates is a man and a stone, Socrates is not a stone.

But 4. If Socrates is not a stone, Socrates is not a man and a stone.

So 5. If Socrates is a man and a stone, Socrates is not a man and a stone.[20]

Conclusion (5) has the logical structure $(p \wedge q) \to \neg(p \wedge q)$ and hence it contradicts ABEL 1. However, Abelard wasn't too much worried by *this* counterexample because he considered step (2) as *not valid*. After all, (2) is a conditional with a *positive* antecedent and a *negative* consequent and hence it only constitutes a *strict*, but, for him, not a "*natural*" implication!

However, Alberic of Paris developed another "embarrassing argument" which could not be rejected in this way:

1.   If Socrates is a man and is not an animal, Socrates is not an animal.

2.   If Socrates is not an animal, Socrates is not a man.

---

[18]Martin (1987, p. 394) stated that Abelard and his followers, the *Nominales*, propounded "the thesis that no conditional can be true in which the antecedent and the consequent are of different quality".

[19]Martin (2004, p. 191) shows that given the laws of conjunction, transitivity, contraposition, and double negation, one obtains $(p \wedge \neg p) \to \neg(p \wedge \neg p)$ as a counterexample to ABEL 1. But a slight variation of this proof without the law of double negation already shows that ABEL 2 is violated anyway. Hence, the law of double negation is not the culprit.

[20]Cf. (Martin, 1986), pp. 569-70. Martin's reconstruction is based on the *Introductiones Montane minores*. A similar argument may also be found in the 13th century *Ars Meliduna*; cf. (Iwakuma, 1993), p. 143.

3. If Socrates is not a man it is not the case that Socrates is a man and an animal.

C* If Socrates is a man and not an animal, it is not the case that Socrates is a man and not an animal. (Martin, 1987), pp. 394–5.

Since conclusion C* has the structure $(p \land \neg q) \to \neg(p \land \neg q)$, it constitutes another counterexample to ABEL 1. Furthermore, the argument does *not* depend on the "locus ab oppositis" because Line 2 is obtained by applying the principle of *contraposition* to the unproblematic conditional 'If Socrates is a man, Socrates is an animal'. Since the proof makes use only of logical laws which Abelard regarded as indispensable, "[...] confronted with this argument Master Peter essentially threw up his hands and granted its necessity" (Martin, 1987, p. 395). In what follows some possibilities to escape "Alberic's trap" shall be investigated.[21]

6   Possible ways out?

Martin (1986, p. 570) remarked somewhat cryptically that "there is some evidence that Abaelard's followers, the *Nominales*, held that only the positive conjunct is entailed by such a conjunction". In (1987), Martin similarly explained "that the *Nominales* adopted the view that from an affirmation and a negation only the affirmative conjunct follows" (p. 397). That means that step 1 of Alberic's argument, i.e., the conditional 'If Socrates is a man and is not an animal, Socrates is not an animal', is *rejected* because in general the conjunction of an "affirmative" proposition p and a "negative" proposition q is supposed to entail only p but not q! Martin (2004, pp. 198-9) described this somewhat strange idea at some greater length:

> They said that 'if Socrates is human and Socrates is not an animal, then Socrates is not an animal' does not hold because a negation is not so powerful (*vehemens*) when joined with an affirmation as it is when it is alone, and something follows from a negation alone which does not follow from it when it is conjoined with an affirmation.

---

[21]The nice expression was coined in (Normore, 1987), p. 205: "[The] confusion about the position of the *Nominales* is just what one would expect if the *Nominales* followed Abelard into Alberic's trap and then had to find their own way out"; I owe this quotation to Courtenay (1993, p. 154).

When I first read this, my spontaneous reaction was to reply:

> This argument represents a typical example of what one nowadays calls a *sophistic solution* of a problem. It is a *possible*, or at least *possibly possible* way out, which, however appears to have been invented entirely *ad hoc* and which is otherwise totally implausible and substantially unfounded. (Lenzen, 2019, p. 550)

Although I still subscribe to this view, it may be helpful to analyse the considerations of the *Nominales* in some more detail.

7   The argument from *Introductiones Montane minores*

After summarizing Alberic's argument as an instance of a proof "that one proposition entails its own negation",[22] the anonymous author of the *Introductiones Montane minores* considers the following objection:

> And they say that it does not follow: '*If Socrates is a man and is not an animal, Socrates is not an animal*', because the negation is not so powerful when it is conjoined with an affirmation as it is *per se*; and something follows from a negation *per se* that doesn't follow from the same proposition when it is conjoined with an affirmation. Thus, from the negation '*Socrates doesn't dispute*' when conjoined with the affirmation '*when Plato is reading*' it doesn't follow '*Socrates doesn't discuss with anybody*', although this follows when the proposition is put forward *per se*.[23]

The latter example, of course, is entirely *correct*. If the proposition 'Socrates doesn't dispute' is *restricted* by adding a qualification like '*when* Plato is reading' ("ubi Plato legit"), then it no longer retains the inferential power of its unrestricted counterpart. But in Alberic's example the situation is very different. The second (negative) conjunct of the antecedent 'Socrates is a man *and* Socrates is not an animal' does *not restrict* the first (affirmative) conjunct. By means of 'and', the contents of both conjuncts are simply "added" but

---

[22]Cf. (De Rijk, 1967), pp. 65–66: "Sicque probari potest quod una propositio infert suam contradictoriam hoc modo: si Socrates est homo et non est animal, Socrates non est animal, et si Socrates non est animal, Socrates non est homo; si Socrates non est homo, non est Socrates homo et non animal, [quare] si est homo et non est animal, non est homo et non est animal."

[23]Cf. (De Rijk, 1967, p. 66), lines 5–12.

not diminished or relativized. This difference was already recognized in the *Introductiones Montane minores*:

> [...] because an affirmation with a negation is understood differently when I say '*Socrates doesn't dispute when Plato is reading*', and differently when I say '*the same [Socrates] is a man and is not an animal*'. For when it is said '*Socrates is a man and is not an animal*', two [things] are predicated and two [propositions] are maintained and understood, namely the sense of an affirmation and the sense of a negation. Therefore, from this [proposition] can be understood everything [that follows] from the affirmation *per se* and everything [that follows] from the negation with the affirmation [.] [But when] it is said '*Socrates doesn't dispute when Plato is reading*', since there is only one proposition which is conjoined from an affirmation and a negation, thus it is not allowed to conclude from it both what follows from the affirmation *per se* and what follows from the negation.[24]

Furthermore, there are serious doubts concerning the consistency of the idea that from the conjunction of an affirmation and a negation only the affirmative conjunct follows. Above all, the presupposition that each proposition might uniquely be classified as *either* affirmative *or* negative appears to be very problematic. E.g., 'Socrates is blind' may be considered as an *affirmation*, yet it is equivalent to the "*negation*" 'Socrates cannot see'. Similarly, both 'Socrates is sick' and 'Socrates is healthy' may be considered as affirmations, but these propositions can equivalently be expressed as 'Socrates is not healthy' and 'Socrates is not sick'. So, what follows from the conjunction 'Socrates is healthy and he is sick'; and what follows from 'Socrates is healthy and he is not healthy', and what from 'Socrates is sick and he is not sick'? Any decision would appear to be entirely arbitrary! Let us therefore consider other possible ways out of "Alberic's trap".

8   The argument of the *Melidunenses*

The followers of Robert of Melun attempted a very surprising solution of the "inconvenientia" brought to light by Alberic's argument. In their chief work *Ars Meliduna*, they put forward the principle that *nothing* follows from

---

[24]Cf. (De Rijk, 1967, p. 66), lines 14–23.

a "false" proposition ("nihil sequitur ex falso"). This principle admits of two interpretations. Taken literally, it says:

EFN   If p is false, then for no q: $(p \rightarrow q)$.

However, some medieval logicians understood the word 'falsum' in the sense of '*necessarily* false'. In this case the principle "only" maintains that nothing follows from an *impossible* antecedent:

EIN   If p is impossible, then for no q: $(p \rightarrow q)$.

The extreme position EFN was severely attacked already by contemporaries. As De Rijk (1967) points out (vol. II/1, p. 281-2), John of Salisbury criticized the dogma of the Melidunenses as conflicting with the views of Aristotle who had explained:

> [...] how true conclusions may be concluded from false premises [...]. He [John of Salisbury] adds somewhat malignantly: this exposition of Aristotle is possibly overlooked by those who contend that nothing follows from what is false.

As was argued in (Lenzen, 2022b), the Melidunenses failed to distinguish with sufficient precision between a *valid* and a *sound* argument. According to our modern understanding, an argument is sound iff it is formally valid *and* it contains only true premises! The question behind principles EIN and ECQ is not, however, whether such inferences are *sound*; what is at issue is only whether they are *formally valid*, or not. Thus, also the Melidunenses eventually conceded that, according to the usual dialectical rules, a *false* (or, indeed, impossible) conclusion like 'Socrates is a stone' can be derived from a *false* (indeed, impossible) premise like 'Socrates is a pearl'. But such an inference may never be *applied* because the attribute 'pearl' cannot be predicated of 'Socrates' since it is opposed to it. For the Melidunenses, logical inferences not only have to be *formally* sound, but also "sound" with respect to *matter*. Furthermore, arguments should serve as *reasons* to help us form an opinion about uncertain states of affairs. But false premises cannot be trusted, so they cannot be used for arguments, "quare ex falso nihil".[25]

As regards the less stringent principle EIN, the Melidunenses believed that the admission of conditionals with impossible antecedents would lead to outright inconsistencies:

---

[25]Cf. (Iwakuma, 1993, pp. 144–145), and the discussion in (Lenzen, 2022b), Section 3.

If one assumes the opposite, it can be proven (1) that a proposition entails its own negation, (2) that two contradictories follow from one and the same proposition, and (3) that a proposition entails another proposition which cannot be true together with it. Each of these conclusions seems to be against the art, for just as nothing can both be [true] and not be [true], so it cannot be the case that two [contradictories follow from] one and the same proposition.[26]

Now it is undeniably true that no proposition q can be both true and not true. Therefore, it is also undeniably true that q and $\neg q$ cannot be *entailed* by one and the same *true* proposition p. But this does *not* mean that q and $\neg q$ cannot simultaneously be entailed by *any proposition at all*. It is not "against the art" that the self-contradictory conjunction $(q \wedge \neg q)$ entails both q and $\neg q$!

9 The way out of the *Porretani*

Martin (1986, pp. 570–571) briefly pointed out that certain logicians "decided that simplification was at fault, since what we assert with such conditionals is that the conjuncts are conjointly sufficient for the consequent but neither alone is sufficient". One year later Martin explained more precisely that he had the followers of Gilbert of Poitiers in mind which are otherwise known as the *Porretani*. Their reason for rejecting the laws of conjunction was put forward in the following passage from the *Compendium Logicae*:

This is because it is a general principle both with regard to consecution and to inference [that it follows] only if the cause of the consequent or the conclusion is preposed to the consequent or conclusion. What indeed is asserted in 'If Socrates is a man and the Seine flows through Paris, then Socrates is an animal'? And what relevance does one of a pair of coupled antecedents have to the consequent when only the other is the cause? Thus in 'If Socrates is a man and an ass, then Socrates is a man', isn't 'Socrates is an ass', which is not a cause, preposed as a cause.[27]

---

[26]Cf. (Iwakuma, 1993), pp. 142–143: "Suscipienti contrarium probari potest (1) ad aliquam propositionem sequi suam contradictoriam, (2) duas contradictorias sequi ad eandem, et (3) ad aliam propositionem sequi aliam quae non potest esse vera cum ista. Quorum quodlibet contra artem esse videtur. Nam quemadmodum nihil potest simul esse et non esse, ita nec ad eandem duae."

[27]This rather free translation of the Latin text in (Ebbesen, Fredborg, & Nielsen, 1983, p. 22) has been adopted from (Martin, 1987), p. 397. It should be noted that Martin's term '*relevance*'

Wolfgang Lenzen

The Porretani thus advocate a strictly causal interpretation of conditionals according to which $(p \to q)$ is true iff the antecedent p is the *cause* of the consequent q. The notion of cause, however, is somewhat unorthodox in so far as a proposition p (or the state of affairs described by p) is apparently accepted as the "cause" of itself. As the second example indicates, Socrates's being a man "causes", and hence implies, that Socrates is a man. Thus, Porretanian implication may be assumed to be reflexive (and probably also transitive). But it fails to satisfy the usual laws of conjunction, because the proposition 'The Seine flows through Paris' is totally *irrelevant* for inferring 'Socrates is an animal' from 'Socrates is a man $\wedge$ The Seine flows through Paris'. Similarly, the counterfactual assumption 'Socrates is an ass' is totally irrelevant for deriving 'Socrates is an animal' from 'Socrates is a man $\wedge$ Socrates is an ass'. In the opinion of the Porretani, whoever holds that if $(p \to q)$, then also $(p \wedge r \to q)$, commits, as they say, the fallacy "*non causa ut causa*": The conjunctive antecedent $(p \wedge r)$ is "preposed" as the *cause* of the consequent q, but the true cause of q is p alone, while $(p \wedge r)$ is a "*not-cause*".

Now the question arises whether the "causal" conception of implication can be made sufficiently precise so as to yield a self-consistent system of logic. Unfortunately, the *Compendium Logicae* is too fragmentary to decide this issue. The only principle that is clearly *accepted* by the Porretani is that of *contraposition*:

CONTRA   If $(p \to q)$, then $(\neg q \to \neg p)$.[28]

Furthermore, they do *not reject* the principle that "of a true conjunction each conjunct is true".[29] But this principle is "normally" interpreted to mean that the falsity of one of the conjuncts, say p, *entails* the falsity of the entire conjunction, i.e., $(\neg p \to \neg(p \wedge q))$. Hence, by applying CONTRA (plus the law of double negation), one would obtain that $(p \wedge q)$ entails p, after all!

---

has no direct counterpart in the original where the crucial sentence just reads: "At quid magis unum antecedentium copulatorum *attinet* ad consequens, cum tantum alterum sit causa?" Note also that the explanatory '[that it follows]' was added not by me but by Martin.

[28]Cf. (Ebbesen et al., 1983), p. 23: "Sed hoc habet maxima artis: si [...] ad unam aliqua sequitur, ad contradictoriam [con]sequentis sequitur contradictoria antecedentis, ut 'si Socrates est homo, Socrates est animal; ergo si non est animal, non est homo". An anonymous referee pointed out that, as explained in § 2.5.3 of (Francez, 2021), another possible way out consists in restricting just this law of contraposition.

[29]Cf. (Ebbesen et al., 1983), p. 21: "Ratio quare dicatur omnis copulative vere utramque partem esse veram. Cum itaque copulativa nomen contrahat a copulando, suam habet veritatem a veritate copulatorum. Si ergo falsa copulet vel falsum vero, cadit a veritate non verum coniungendo vero; ad hoc enim ut sit vera oportet ut vera coniungat."

As Martin remarked in (1987, pp. 397–398) the reservations of the Porretani concern "exactly the point made by Everett Nelson in his account for the intensional relationship holding between the antecedent and consequent of a true conditional." Let us therefore jump into the 20th century and see whether Nelson succeeded in finding a way out of "Alberic's trap".

10   Nelson's theory of "intensional relations"

Nelson's paper was written at a time when modal logic was still in its infancy. Nelson wanted to show:

  (i) that purely extensional logics as developed, e.g., in *Russell's Introduction to Mathematical Philosophy*, are insufficient to capture the logical behaviour of the relation of the inconsistency of two propositions p, q, and of the relation of entailment between p and q;

 (ii) but that the intensional system of C. I. Lewis's *Survey of Symbolic Logic* is inadequate because it gives rise to the "paradoxes of strict implication".

Nelson accepted Lewis's idea of determining the truth-conditions of $(p \rightarrow q)$ in terms of the "inconsistency" of $(p \wedge \neg q)$;[30] yet the resulting implication, which Nelson calls 'entailment', differs from Lewis's strict implication because it is based on a different conception of the (in)consistency of two propositions. Lewis presupposes the "normal" interpretation according to which p and q are incompatible iff $(p \wedge q)$ is impossible. As a consequence of this definition, it follows that a self-contradictory proposition like $(p \wedge \neg p)$ is incompatible with itself. This, however, Nelson (1930) thinks, is *wrong*:

> Though it may be that a part of a proposition is inconsistent with another part, the proposition as a whole is not inconsistent with itself as a whole. For example, in regard to the proposition 'All men are mortal and some men are not mortal', the two component propositions are inconsistent with each other, but the whole compound is not inconsistent with itself. (p. 447)

Nelson clearly saw that, in order to avoid the "paradoxes of strict implication", the usual laws of conjunction have to be abandoned.

---

[30]Cf. (Nelson, 1930), pp. 444–445: "The propositional function 'p entails q' means that p is inconsistent with the propositional function that is the proper contradictory of q (i.e., with $\neg q$)".

Wolfgang Lenzen

Conjunction: $p \,\&\, q$. I do not take $p \,\&\, q$ to mean 'p is true and q is true', but simply 'p and q', which is a unit or whole, not simply an aggregate, and expresses the joint force of p and q. I say that $p \,\&\, q$ is not a simple aggregate composed of p and q because, for example, in asserting that $p \,\&\, q$ entails r, I am not asserting something simply about p and about q, but about them taken together, i.e., their *joint* force. $p \,\&\, q$ does not entail r unless both p and q *function together* in entailing r. It is this functioning together or joint force that gives to conjunction a unity which a mere aggregate or collection does not have. And this unity of conjunction is in a sense relational; *e.g.*, the propositions 'All men are mortal' and 'Socrates is a man' form a conjunction in relation to 'Socrates is mortal', for 'Socrates is mortal' expresses their joint force; but they do not form such a conjunction in relation to 'Socrates is a man'. However, if $p[\wedge]q$ alone be asserted, then the mere aggregate of p and q is asserted; i.e., p is asserted and q is asserted. In such a case the unity of conjunction is lost, and the mere assertion of each component remains. (Nelson, 1930, p. 444)

Nelson here introduces an interesting distinction between two possible interpretations of 'conjunction': The "joint" view and the "aggregate" view. These two brands are best distinguished by means of different symbols. While Nelson symbolizes a conjunction of p and q by mere juxtaposition as '$p\,q$', let us continue to use '$\wedge$' for the "aggregational", truth-functional conjunction, while the "joint" conjunction shall be symbolized by '$\&$'.[31] Now, a closer investigation of the *logic* of the operator of "joint" conjunction appears to be a rewarding task, but it clearly falls beyond the scope of this paper. Suffice it to point out that, whatever the proper laws of the "joint" conjunction eventually turn out to be, they do not affect the validity of the laws of "aggregate" conjunction which consist of the principles of conjunction-*elimination*,

CONJ 1 $(p \wedge q) \to p$

CONJ 2 $(p \wedge q) \to q$,

---

[31] As a matter of fact, in the foregoing quotation most occurrences of Nelson's '$p\,q$' have already been replaced by '$p \,\&\, q$'. Only the last occurrence (in the third last line) was replaced by '$p[\wedge]q$' in order to indicate that what is at issue there is not the assertion of a "joint" conjunction but that of an "aggregational" conjunction.

70

plus the inference of conjunction-*introduction*:

CONJ 3 $p, q \Rightarrow (p \wedge q)$.

According to Nelson (1930), the "joint" conjunction does not satisfy these laws:

> Naturally, in view of the fact that a conjunction must function as a unity, it cannot be asserted that the conjunction of p and q *entails* p, for q may be totally irrelevant to [...] p, in which case p *and* q do not entail p, but it is only p that entails p. I can see no reason for saying that p and q entail p, when p alone does and q is irrelevant, and hence does not function as a premise in the entailing. (p. 447)

Now even it is taken for granted that, for reasons of *relevance*, the sophisticated operator of "joint" conjunction, '&', fails to satisfy CONJ 1 (and thus, for reasons of *symmetry*, also fails to satisfy CONJ 2), it still remains a fact that the less sophisticated, truth-functional operator '$\wedge$' *does* satisfy these laws. Thus, also Nelson (1930) came to admit:

> I do not deny that from 'p is true and q is true' we can pass to 'p is true'. All I deny is that such a passage is in virtue of an entailment relation holding between 'p is true and q is true' and 'p is true'. (p. 448)

The mitigating expression of '*passing*' from the truth of $p \wedge q$ to the truth of the conjuncts evidently means nothing else but that if $p \wedge q$ is true, then p must be true, and q must be true, as well. But this is just what principles CONJ 1, 2 are saying! So even if the corresponding inferences from $p \& q$ to p, or to q, are *not* valid, nevertheless $p \wedge q$ "entails" p, and "entails" q, in the sense that the falsity of p, and the falsity of q, is incompatible with the truth of $p \wedge q$.

Estrada-González and Ramírez-Cámara (2020) argued that Nelson's rejection of the laws of conjunction (or, as they call it, the principle of "simplification") shows a possible way out of the problems created by Alberic's "embarrassing argument". But the authors admitted:

> The obvious objection to the Nelsonian escape from Alberic's argument is that rejecting Simplification has a higher theoretical cost than rejecting the unrestricted validity of the connexive principles, for the latter are much less intuitive than Simplification. (p. 101)

Of course, it is a rather subjective matter which principle one considers as more intuitive than the other. One need not necessarily share the opinion of Routley et al. who believe that "rejecting Simplification is 'the wrong choice', is 'too drastic', and would have a 'devastating effect' on 'the practice of reasoning'", so that deductive "reasoning would become unworkable'".[32] Furthermore, as Estrada-González and Ramírez-Cámara (2020) point out, the rejection of simplification must not be treated as a matter of yes or no but may proceed in degrees. Instead of maintaining that the laws of conjunctions *never* hold, one might assume that (i) "Simplification fails if and only if the conjuncts are incompatible" or that (ii) "Simplification fails if and only if the conjuncts are not minimally relevant to each other (namely, if they do not share any propositional variables)" (pp. 101–102). However, as was noted already at the beginning of Section 9, the "relevant" approach (ii) is unapt to cope with the problem that $((p \land \neg p) \to p)$ together with $((p \land \neg p) \to \neg p)$ constitutes a counterexample to ABEL 2.[33]

Estrada-González & Ramírez-Cámara consider Nelson's approach as rather obscure.[34] At any rate, restricting the laws of *conjunction* will not be of much help because, as Nelson clearly recognized, the laws of *disjunction*

DISJ 1 $p \to (p \lor q)$

DISJ 2 $q \to (p \lor q)$

allow to construct similar counterexamples. As was discovered already in the 13th century by Robert Kilwardby:

> [...] a disjunctive follows from either of its parts, and in a natural inference. Hence it follows 'If you are sitting you are sitting or you are not sitting' and 'If you are not sitting, you are sitting or you are not sitting'. And thus one and the same thing

[32]Cf. (Estrada-González & Ramírez-Cámara, 2020), p. 101. The authors take the quotation from Chapter 2 of (Routley, Plumwood, Meyer, & Brady, 1982) and they mention that Wansing and Skurt (2018) want to avoid "the highly unintuitive failure of Simplification".

[33]Furthermore, Estrada-González and Ramírez-Cámara (2020) think that the "consistency" approach (i) alone also is not apt to cope with the counterexamples. Therefore, they suggest to combine (i) and (ii) by requiring that (iii) "Simplification fails if and only if the conjuncts do not share literals".

[34]Cf. (Estrada-González & Ramírez-Cámara, 2020), p. 109: "According to Nelson, a conditional is true if and only if the negation of the consequent is incompatible with the antecedent. [...] Nelson's ideas are obscure, without enough formalization to facilitate understanding of what he is proposing".

Abelard and the Development of Connexive Logic

follows in a natural inference, and thus of necessity, from the same thing's being so and not being so.[35]

In order to escape this dilemma, Nelson (1930) introduced, besides the usual operator '$\vee$', another operator of "intensional logical sum" which he symbolized by a bold, capital '$\mathbf{V}$':

> Intensional logical sum: $p \mathbf{V} q. = . - p/ - q.$ '$p \mathbf{V} q$' may be read 'either p or q', but it holds only if there is a necessary connexion between p and q [...]. It must not be confused with '$p \vee q$' which is merely factual. (p. 446)

This "intensional" disjunction is kind of a *necessary* disjunction, which in the framework of standard modal logic might be defined by the condition that $\neg p$ *strictly* implies q, i.e., $\neg\Diamond(\neg p \wedge \neg q)$, or simply $\Box(p \vee q)$! Hence the counterparts of the standard laws for "material" disjunction, DISJ 1, 2 *fail to hold* for "strict" disjunction. E.g., the mere *truth* of the antecedent, or premise, p, cannot guarantee that the consequent, or conclusion, $(p \vee q)$ is *necessarily* true. Such an inference is warranted only if the premise is strengthened to the requirement that p (or q) be *necessary*:

DISJ 3 $p \rightarrow (p \mathbf{V} q)$

DISJ 2 $q \rightarrow (p \mathbf{V} q)$.

This does not, however, in any way affect the validity of the laws of "material" disjunction, DISJ 1, 2. Thus, Nelson had to admit that "of course, if p has truth, then 'p or q' has truth", or, as one might also say in Nelson's peculiar terminology, from 'p is true' we can pass to 'p or q' is true'. Yet Nelson (1930) insists that a corresponding *entailment* would not be valid:

> [...] 'p entails p or q' cannot be asserted on logical grounds, because from an analysis of p we cannot derive the propositional function 'p or q' where q is a variable standing for just any other propositional function whatsoever. (p. 448).

It is interesting to note that in this passage Nelson fails to make clear what kind of disjunction is at issue. He simply speaks of 'p or q', leaving it open whether he has '$p \mathbf{V} q$' or '$p \vee q$' in mind. Of course, as everyone has to grant,

---

[35]Cf. (Thom & Scott, 2015), p. 1141; for a detailed discussion of Kilwardby's treatment of Aristotle's Theses cf. (Lenzen, 2020b).

a "strict" or *necessary* disjunction is not entailed by either of its disjuncts; but the truth of a "*material*" disjunction *follows from* the truth of either disjunct in the sense that the falsity of $p \lor q$ is *incompatible* with the truth of p (and similarly, with the truth of q).

When Nelson remarks that from the "analysis of p we cannot derive [...] 'p or q'" he somehow misses the point. We don't have to analyse the *propositions* p or q or 'p or q' in order to find out whether one entails the other or not. All we need to do is to analyse the *logical operator* 'or', which, of course, admits of many different interpretations, e.g., "inclusive" vs. "exclusive" disjunction, "material" vs. "strict" disjunction, etc. These different operators will normally be characterized by different logical laws. While the properties of inclusive disjunction '$\lor$' are determined by the well-known truth-table which "defines" '$p \lor q$' to be false if and only if both p and q are false, it would have been Nelson's task to explain the properties of his "intensional logical sum" '$p \mathbf{V} q$' in more detail.[36] Unfortunately, he remains largely silent on this point.

To conclude this section, let us put Nelson's ideas once again in a broader historical context. When Nelson wrote his paper, the general situation was very different from now. It was yet a long way to go to the invention of possible-worlds semantics; and a much longer way to the development of logics of, e.g., counterfactual *conditionals*, the semantic analysis of which reaches far beyond Kripke semantics. We now know that, besides Lewis's calculi of strict implication, there exists a (potentially) *infinite* variety of different modal logics, and hence, besides the simple, truth-functional operator of *material* implication, also a (potentially) infinite variety of *strict* implications. In 1930, it was important to emphasize that the (in)consistency between two propositions is an "intensional" relation; and it was also important to discuss the so-called paradoxes of strict implication to which the Lewisian operator(s) gave rise. During the past decades, many attempts have been made to avoid these "paradoxes" by constructing non-classical systems of relevance logic, paraconsistent logic, and connexive logic. Some of Nelson's ideas anticipate the development of these logics, and it would seem to be a rewarding task to investigate whether the concept of a "joint" conjunction '&' and of a "strong" disjunction '$\mathbf{V}$' can be rendered sufficiently precise so as to yield a full-fledged calculus of *relevance logic*. This task, however, clearly lies beyond the scope of this paper.

---

[36] In particular, it would be interesting to know how Nelson's '&' and '$\mathbf{V}$' behave with respect to *negation*. Are there any De Morgan like laws which regulate their interrelations?

Let us conclude this section by noting that, at any rate, Nelson failed to provide a convincing *explanation* why a self-contradictory proposition like 'All men are mortal *and* some men are not mortal' must not be considered as *inconsistent with itself*. Of course, it is true that the inconsistency of two propositions "normally" results from the fact that one proposition, say p, somehow contradicts the *other* proposition, q. Such a contradiction obtains in particular if, in the terminology of traditional logic, either p and q are *contradictories* (like 'All men are mortal' and 'Not all men are mortal'), or *contraries* (like 'All men are mortal' and 'All men are immortal'). But most contemporary logicians have no qualms to consider a *logically false* proposition like '$p \wedge \neg p$' as *self-contradictory* or *self-inconsistent*! After all, the *conjunction* $(p \wedge \neg p)$ contradicts itself in the sense that if, e.g., the first conjunct, p, would be true, then the second, $\neg p$, would not be true and hence the entire conjunction would not be true either.

11   Conclusion

As Martin noted in (1986, p. 571) another "way to handle the difficulties of connexive logic is simply to take condition [STRICT] as the criterion of truth for a conditional and to give up the connexive principles or else say that they do not apply to the paradoxical cases." Similarly, in their exposition of the history of the principle "Ex contradictorio quodlibet", Priest, Tanaka, and Weber (2018) explained:

> Abelard's position was shown to face a difficulty by Alberic of Paris in the 1130s. [...]. But one way to handle the difficulty is to reject the connexive principle. This approach, which has become most influential, was accepted by the followers of Adam Balsham or Parvipontanus [...]. The Parvipontanians embraced the truth-preservation account of consequences and the 'paradoxes' that are associated with it.

Let it be emphasized that the Parvipontanian solution does not imply to give up the entirely plausible requirement of *humble* connexivism which was accepted by each medieval logician. The idea of the Parvipontanians only means to give up *hardcore* connexivism which wants to extend connexivity to the "paradoxical" cases of an impossible antecedent and a necessary conse-quent. There may perhaps be good reasons for developing systems of hard-core connexive logic,[37] but, as was stressed in (Lenzen, 2022a), p. 553:

> Whatever the motives may be which guide contemporary logicians in building calculi which satisfy such "hardcore" conditions, they should at least stop claiming that *their* logics are elaborations of ideas that can be traced back to Aristotle, Boethius, Abelard, Kilwardby, etc.

References

Abaelardus, P. (1970). *Dialectica* (L. M. de Rijk, Ed.). Assen: Van Gorcum.

Courtenay, W. J. (1993). Nominales and rules of inference. In K. Jacobi (Ed.), (pp. 153–160). Leiden: Brill.

De Rijk, L. M. (Ed.). (1967). *Logica Modernorum – A Contribution to the History of Early Terminist Logic, 2 vols.* Assen: Van Gorcum.

Ebbesen, S., Fredborg, K. M., & Nielsen, L. O. (Eds.). (1983). Compendium logicae Porretanum ex codice Oxoniensi Collegii Corpori Christi 250. *Cahiers de l'Institut du Moyen-Âge Grec et Latin, 46*, 1–93.

Estrada-González, L., & Ramírez-Cámara, E. (2020). A Nelsonian response to 'the most embarrassing of all twelfth-century arguments'. *History and Philosophy of Logic, 41*(1), 101–113.

Francez, N. (2021). *A View of Connexive Logics.* London: College Publications.

Geyer, B. (Ed.). (1919–1933). *Peter Abaelards Philosophische Schriften. Beiträge zur Geschichte der Philosophie und Theologie des Mittelalters* (Vols. 21, Hefte 1–4). Münster: Aschendorf.

Iwakuma, Y. (1993). Parvipontani's Thesis Ex Impossibili Quidlibet Sequitur: Comments on the Sources of the Thesis from the Twelfth Century. In K. Jacobi (Ed.), (pp. 123–151). Leiden: Brill.

Jacobi, K. (Ed.). (1993). *Argumentationstheorie – Scholastische Forschungen zu den logischen und semantischen Regeln korrekten Folgerns.* Leiden: Brill.

Kapsner, A. (2019). Humble connexivity. *Logic and Logical Philosophy, 28*(3), 513–536.

---

[37]One possible reason was stated in (Wansing & Skurt, 2018, p. 484), as follows: "[…] in the area of computer science, and non-monotonic reasoning in particular, the idea that contradictions should not have any inferential power is not new and has been suggested with no reference to the cancellation model of negation".

King, P., & Arlig, A. (2018). Peter Abelard. In E. N. Zalta (Ed.), *The Stanford Encyclopedia of Philosophy* (Fall 2018 ed.).

Kneale, W., & Kneale, M. (1962). *The Development of Logic*. Oxford: Clarendon.

Lenzen, W. (2019). Leibniz's laws of consistency and the philosophical foundations of connexive logic. *Logic and Logical Philosophy*, *28*(3), 537–551.

Lenzen, W. (2020a). A critical examination of the historical origins of connexive logic. *History and Philosophy of Logic*, *41*(1), 16–35.

Lenzen, W. (2020b). Kilwardby's 55th lesson. *Logic and Logical Philosophy*, *29*(4), 485–504.

Lenzen, W. (2021). *Abaelards Logik*. Paderborn: Brill/mentis.

Lenzen, W. (2022a). Rewriting the history of connexive logic. *Journal of Philosophical Logic*, *51*(3), 525–553.

Lenzen, W. (2022b). What follows from the impossible: Everything or nothing? (An interpretation of the 'Avranches text' and the Ars Meliduna). *History and Philosophy of Logic*, *43*(4), 309–331.

Martin, C. (1986). Williams' machine. *The Journal of Philosophy*, *83*(10), 564–572.

Martin, C. (1987). Embarrassing arguments and surprising conclusions in the development of theories of the conditional in the twelfth century. In J. Jolivet & A. de Libera (Eds.), *Gilbert de Poitiers et ses Contemporains* (pp. 377–400). Napoli: Bibliopolis.

Martin, C. (2004). Logic. In J. Brower & K. Guilfoy (Eds.), *The Cambridge Companion to Abelard* (pp. 158–199). Cambridge: Cambridge University Press.

McCall, S. (2012). A history of connexivity. In D. M. Gabbay, F. J. Pelletier, & J. Woods (Eds.), *Handbook of the History of Logic, Vol. 11* (pp. 415–449). Amsterdam: Elsevier.

Nelson, E. (1930). Intensional relations. *Mind*, *39*(156), 440–453.

Normore, C. (1987). The tradition of mediaeval nominalism. In *Studies in Medieval Philosophy* (pp. 201–217). Washington: The Catholic University of America Press.

Pizzi, C. (1977). Boethius' thesis and conditional logic. *Journal of Philosophical Logic*, *6*(1), 283–302.

Pizzi, C., & Williamson, T. (1997). Strong Boethius' thesis and consequential implication. *Journal of Philosophical Logic*, *26*(5), 569–588.

Priest, G., Tanaka, K., & Weber, Z. (2018). Paraconsistent logic. In E. N. Zalta (Ed.), *The Stanford Encyclopedia of Philosophy* (Summer 2018 ed.).

Routley, R., Plumwood, V., Meyer, R. K., & Brady, R. T. (1982). *Relevant Logics and Their Rivals* (Vol. 1). Atascadero: Ridgeview.

Thom, P., & Scott, J. (Eds.). (2015). *Robert Kilwardby, Notule libri Priorum.* Oxford: Oxford University Press.

Wansing, H. (2020). Connexive logic. In E. N. Zalta (Ed.), *The Stanford Encyclopedia of Philosophy* (Spring 2020 ed.).

Wansing, H., & Skurt, D. (2018). Negation as cancellation, connexive logic, and qLPm. *Australasian Journal of Logic*, *15*(2), 476–488.

Wolfgang Lenzen
University of Osnabrück, Department of Philosophy
Germany
E-mail: `lenzen@uos.de`

Towards a Formal Analysis of Semantic Pollution of Proof Systems

ROBIN MARTINOT[1]

Abstract: This paper concerns itself with what it means to be a 'good' proof system, and we argue that a satisfying answer should incorporate the goal that a proof system is intended to have. We outline the main goals of proof systems, and claim that properties of a proof system should be considered 'good' if they contribute to achieving the particular goal of the proof system. Our next interest is to characterize the notion known in the literature as 'semantic pollution'. We suggest that the value of semantic influence in proof systems is goal-dependent, and that semantic influence can become pollution if it stands in the way of formalizing informal reasoning. In an attempt to characterize semantic pollution more systematically, we consider categoricity and the existence of a semantic translation of proof rules as candidate properties, but conclude that neither is sufficient. Finally, we suggest that meta-theoretic considerations might induce a separate type of semantic pollution.

Keywords: philosophy of proof theory, formalization, semantic pollution

1 Introduction

One of the occupations of philosophy of logic is finding criteria for when formal systems are 'good'. In this paper we will focus on criteria for proof systems. Proof theory is a field that was originally strongly motivated by the aim to provide a consistent foundation of mathematics, but has turned out to provide applications in many areas, such as computer science, AI, logics for non-mathematical reasoning, and philosophy. Thus, the reasons for designing proof systems have broadened, and many different types of proof systems are sprouting up.

However, a consistent way to decide whether the properties of proof systems are desirable or not is lacking, and we think more guidance is welcome. The first part of this paper proposes that a property of a proof system

[1] I am grateful to Rosalie Iemhoff and Luca Incurvati for fruitful discussions, and acknowledge support from the Netherlands Organisation for Scientific Research under grant 639.073.807.

should be judged relative to the intended goal of the proof system. That is, analogously to Maddy (2019)'s characterization of the jobs of foundations for mathematics, we can ask "what do we want a proof system to do?". We outline what we believe to be the main goals of proof systems (Provability, Applicability, Formalization, Normativity and Inferentialism), and suggest to consider properties of proof systems 'good' if they contribute to the goal of their proof system.

The second part of the paper focuses on the value of semantic influence in proof systems, and in particular on the phenomenon of semantic pollution. It has been mainly proof systems that incorporate Kripke semantics into the syntax (see e.g. Negri, 2005) that have been considered semantically polluted. The notion of semantic pollution lacks an explanatory formal characterization, and we consider (and reject) some candidates for such an analysis here.

2 What do we want a proof system to do?

This section describes the main goals with which proof systems are commonly designed. We incorporate previous literature into our distinction between five goals: Provability, Applicability, Formalization, Normativity and Inferentialism. Formal properties of proof systems that turn out fruitful for the achievement of these goals, should then be considered desirable. As properties can turn out to be beneficial for one goal, but harmful for another, it is important to clearly establish what goal is adopted. We shortly discuss each purpose separately, distinguishing between formal and philosophical goals.

Formal goals.   A proof system has a formal purpose if it is designed for the sake of establishing (or contributing to establishing) some technical result. The most direct formal purpose of a proof system is the following.

(Goal 1) Provability – A proof system should ensure provability of a set of theorems.

When the most important task of a proof system is to "characterize what results [...] follow from certain axioms [— i]n other words: to investigate the proof-theoretic strength of particular formal systems" (Buss, 1998), its goal is Provability. Here, one is simply interested in what statements are and are not provable in a certain theory. Most commonly, this concerns mathematical provability, but we do not exclude a similar goal for formal logical reasoning.

We note that this goal assumes that the proof system is able to 'accurately' measure proof-theoretic strength, and so that the proof system formalizes a proper notion of proof. This relates to Avron (1996)'s requirement on proof-theoretic frameworks that they possess "the proof-theoretic nature and the expected generality". Thus, while Provability aims to tell us something about the proof-theoretic consequences of a certain formalized theory, the systems used also implicitly clarify what we are willing to call a 'proof' system (for example, which syntactic ingredients, balance between axioms and inference rules, or form of the rules may be used).

However, a proof system can also aim to achieve other formal interests. We capture this by introducing a second goal of designing a proof system.

(Goal 2) Applicability – A proof system should lead to new results in proof theory or other areas of research.

There are many instances of proof systems that are not truly designed to ensure provability of certain theorems, but rather to establish a property that is valuable for obtaining another technical result. We distinguish between applicability to results within proof theory, and to results outside of proof theory (we recognize that this covers much ground, and that this goal could be split up into subgoals for more nuanced distinctions).

Within proof theory, a proof system may be designed for the "study of proofs as objects of independent interest" (Buss, 1998). We might want to find formal proofs with certain syntactic properties, such as normality or cut elimination, or properties concerning the structure of proofs. But proof systems can provide new results within a wide range of other research areas. Applications in computer science include proof generation and proof checking, while new results in related mathematical areas like category theory may also be obtained. More logical results include soundness and completeness (e.g. Negri, 2011 for modal logic), obtaining the interpolation property (see e.g. Fitting & Kuznets, 2015), or even categoricity results with respect to a semantics (see e.g. Rumfitt, 1997). Avron (1996) describes the desideratum of proof systems to be able to handle a great diversity of logics, to improve our understanding of them. Finally, related is also the desire that, in order to encourage possible applications, the structures used in a proof system "should not be too complicated" (Avron, 1996). The list of applications undoubtedly runs much longer than what is described here, but we trust that this goal is sufficiently clear.

We now turn to more philosophical purposes of designing a proof system.

Philosophical goals.   Proof systems are also regularly designed in order to achieve a philosophical value. We identify three main ones.

(Goal 3) Formalization – A proof system should reflect (a property of) an informal way of reasoning.

This is a relevant requirement for a proof system, if one is interested to understand the relation between informal and formal reasoning better. It is related to Steinberger (2011)'s Principle of Answerability of proof systems, which says that "only such deductive systems are permissible as can be seen to be suitably connected to our ordinary inferential practices". Steinberger argues, for instance, that employing multiple-conclusion proof systems do not satisfy this principle.

But again, there is a broad spectrum of properties of informal reasoning that can fall under this notion. We can distinguish between proof systems that aim to formalize entire ways of informal reasoning, and those that focus only on properties of informal reasoning. For example, some proof systems are intended to formalize deontic reasoning (e.g. van der Torre & Villata, 2014), or to focus on avoiding paradoxes such as logical omniscience. Then there are philosophical properties of proof such as explanatoriness, grounding, purity and simplicity, that have a firm intuitive basis, but often lack easy expressibility in formal proof systems. Poggiolesi (2020) develops a proof system that formalizes the instances of informal explanatory proofs that are capturable by formal systems, a clear philosophical goal, and Arana (2009) checks (although the answer is negative) whether proof systems that have the subformula property are a suitable formalization of 'pure' proofs. On a more fundamental level, the style of proof system to begin with is thought to influence suitability of formalization. There, natural deduction is often mentioned as "[intending] to capture the way [we] actually reason" (Bimbó, 2014), more than systems such as sequent or Hilbert calculi. Whether this is actually the case, seems to deserve closer inspection. We also note that simplicity of design is again relevant here, as complex formalizations will often leave behind the intuition of informal reasoning.

A related, but quite different, goal of proof systems should be separated from Formalization.

Towards a Formal Analysis of Semantic Pollution of Proof Systems

(Goal 4) Normativity – A proof system should tell us how to reason.

Taking Formalization a step further, means that we trust the proof system to be a reasoning system that we can (and should) look to for improving our arguments (see e.g. Tosatto, Boella, van der Torre, & Villata, 2012). This area has potential applications in, for example, legal reasoning.

Our last goal, also philosophical in nature, has gained more recent attention in the literature.

(Goal 5) Inferentialism – A proof system should assign meaning to the logical connectives.

Inferentialism says that "the meaning or significance of logical constants is a matter of the inferential rules, or the rules of proof, which govern them" (Peregrin, 2012). A well-known worry for inferentialists is that we can devise rules for a connective *'tonk'*, that allow us to prove anything, and that render the connective meaningless. The common conclusion is that there should be constraints on inference rules to make them suitable for inferentialism. Wansing (1994) discusses various desirable properties of meaning-assigning proof systems, such as Separation, which requires that an inference rule should not introduce more than one connective. He proposes the display calculus as suitable for assigning meaning to logical constants, as it satisfies these conditions. Similarly to the argument for Formalization, Steinberger (2011) says that multiple-conclusion sequent systems are not suitable for inferentialism, because they do not reflect our inferential practices. Hence, there are more philosophical reasons as well as more formal properties for judging compatibility of a proof system with inferentialism.

In short, the purpose that a proof system is intended to serve can differ greatly – but knowledge of its goal is important to be able to decide what (formal) properties are desirable for it. This will also give more insight into the different reasons behind the use of properties, instead of their getting a general (imprecise) image of being 'good' or 'bad'. For example, Poggiolesi (2009, 2010) relevantly lists a number of desirable properties such as the subformula property, invertibility of rules, explicitness of rules, and so on — that would still benefit from a clearer embedding into what a proof system is intended to achieve. Furthermore, there exist many contemporary generalizations of the sequent calculus (hypersequents, nested and labeled sequents, and display sequents) that we think deserve more goal-directed attention.

3 The case of semantic pollution

We are now interested in understanding the phenomenon of semantic pollution of proof systems better. For one, what drives the term pollution in the labeled calculus? We suggest we can view the labeled calculus as a proof system failing to achieve Formalization. Here, semantic influence that is 'artificial' in that it is not reflective of informal reasoning, becomes pollution. Second, we wonder whether the case of labeled calculi can be generalized as being an instance of an overarching formal property of proof systems that identifies semantic pollution. We mention some previous characterizations of semantic pollution in Section 3.3, while Section 4 considers the suitability of several other properties.

3.1 What is semantic pollution?

Intuitively, semantic pollution of a proof system is the idea that a proof system shows an 'unnaturally' strong connection to a (model-theoretic) semantics, so that semantic influences can be recognized in the proof system. Such influences can be connections to a specific type or style of semantics, or to a model in particular. The notion traces back to Avron (1996), who claims that "[b]ecause of the proof-theoretical nature and the expected generality, the [proof-theoretical] framework should be independent of any particular semantics. One should not be able to guess, just from the form of the structures which are used, the intended semantics of a given proof system". This suggests that model-theoretic semantics is really something separate from what proof theory intends to do, and its influence can 'pollute' a proof system.

More recent analyses of semantic pollution focus on the proliferation of adapted and generalized versions of sequent calculi. In particular, labeled calculi for modal and intuitionistic logic (see e.g. Negri, 2005) are associated with the phenomenon. These calculi incorporate Kripke semantics into the proof system, so that every formula is labeled by a world variable w and a forcing relation ':'. Additionally, relational atoms xRy may occur in the sequents, which are motivated by the accessibility relation between worlds. See for example the rules for $\Box$ of the system **G3K** (Negri, 2005):

$$\frac{y : A, x : \Box A, xRy, \Gamma \Rightarrow \Delta}{x : \Box A, xRy, \Gamma \Rightarrow \Delta} \; L\Box \qquad \frac{xRy, \Gamma \Rightarrow \Delta, y : A}{\Gamma \Rightarrow \Delta, x : \Box A} \; R\Box$$

Although systems that incorporate Kripke semantics are most well-known, Negri (2016) also designs a proof system that incorporates neighborhood semantics; and she describes a general method for moving from a model-theoretic semantics to proof-theoretic inference rules.

3.2 Semantic pollution as failing to achieve Formalization

In order to better understand semantic pollution conceptually, we here consider its effect on goals of proof systems. Generally, it is one of the main tasks of logical systems to provide a syntax and semantics that naturally fit together. Thus, we can value "the ability of a syntactic system to adequately reflect semantics, on a par with more familiar properties such as soundness and completeness" (Bonnay & Westerståhl, 2016). If a proof system is far removed from a semantics, perhaps this suggests that we find ourselves outside of what we really mean by formal provability. Given a logic or theory, we might then expect that a proof-theoretic and model-theoretic style of reasoning show compatibility, instead of connecting two systems of which we do not see intuitively why they should relate, in which case the relation could just be a coincidence. On the other hand, we also value soundness and completeness results that are (conceptually) non-trivial. That is, we would like the syntactic side and the semantic side to be considered truly independent from each other, instead of being artificially similar and exposing only superficial connections between each other. Thus, there is a certain 'natural' connection as well as distance between proof theory and model theory that is desirable. A forceful 'plugging in' of semantics into syntax seems to let the connection between proof theory and model theory be too easily achieved, and to disregard the distance. Taking labeled calculi as the main example of semantic pollution, we consider whether semantic influence in proof systems becoming unnatural (and so pollution) has any connection to the goals of proof systems.

Semantic influence in a proof system can prove highly useful for formal goals such as Provability and Applicability. For example, Negri (2011) notes of labeled calculi that "their expressive power, analyticity, applicability to proof search, the possibility to obtain direct completeness proofs without artificial Henkin-set constructions, and their use in the solution of problems that usually involve complex model-theoretic constructions such as negative results in correspondence theory and modal embeddings among different logics" are advantages of including (in this case Kripke) semantics in a proof system. And "the transparent semantic motivation behind the rules makes them intuitive and allows a direct completeness proof" (Negri, 2017).

Robin Martinot

These are all instances of formal results that a proof system can contribute to by adding the labeling structure. Semantic influences may here, too, feel 'unnatural', but we see that this is not because they prevent achievement of Provability and Applicability.

The compatibility of semantic influence with Inferentialism, furthermore, is debatable. For this, it needs to be settled what sources inference rules may extract meaning from for their connectives. If proof rules are meant to reflect inferences as they occur in deduction in practice, Formalization may be very connected to this goal (as Steinberger (2011) suggests), and similar problems may hold for its achievement. If it does not matter so much what types of reasoning proof rules are inspired by, as long as connectives gain their meaning through some mode of reasoning, then semantic reasoning may well be suitable. And if, as Read (2015) proposes, properties like harmony are decisive for use by inferentialists, then labeled calculi remain a candidate for achieving this goal. We leave this discussion up to further research.

Instead, we think there is reason to believe that semantic influence becomes polluting when it prevents achievement of Formalization. We will illustrate this by the case of labeled calculi. In general, if a model-theoretic structure is far removed from how we actually reason, semantic influence can make a proof system less reflective of practical deduction, and more distance between the two sides is needed for Formalization. Instead, if anything that occurs in the proof system should correspond to some relevant notion of informal reasoning, a proof system has a certain neutrality with respect to the precise formal structure of the semantics. Whatever model-theoretic structures turn out to match the proof system, they should be analyzed independently from the proof theory, which should be open to all kinds of semantic counterparts. Enforcing this strictly could mean that a proof system should not even be able to mention (and show bias to) the specific formal structures that occur on the model-theoretic side.

Specific to labeled calculi like **G3K**, criticisms against its semantic influences are often only mentioned superficially or waved away, for instance because the rules still "reflect the intuitive meaning of the logical constants" (Negri, 2011). We maintain, however, that criticisms need to be taken seriously if the calculus is intended to achieve Formalization. Compatibility of the labeled calculus with this goal requires more work. For instance, to obtain Formalization, we need an argument that the extra syntax that occurs in a proof system (intended to refer to world variables w, the forcing relation

$\vDash$, and the accessibility relation R from Kripke semantics) has a natural role in modal reasoning in practice, which seems not at all clear.

We may use phrases like "it is necessarily true, because things could not have been otherwise", but the proof-theoretic machinery of symbols for possible worlds, a forcing relation, and the accessibility relation are very explicit compared to what we actually express linguistically. Furthermore, the labeled calculus uses the term xRy only for modal inference steps, but labels $x : \varphi$ are used for every formula in non-modal arguments. Especially in the latter case, the formal machinery is essentially redundant.

A different way that labeled calculi could strive to satisfy Formalization is implicitly. It may be that we use possible worlds, truth at a possible world, and an accessibility relation only indirectly, perhaps even only subconsciously, in informal reasoning. These elements cannot consistently be found in natural language itself, but they may explain what we are 'actually' doing when we make modal arguments. However, Menzel (1990) gives reason to be cautious with this idea. Concerning Kripke semantics that includes haecceities (essences), he says that "the lack of any overt modalities in the truth conditions on the right-hand side might foster an illusion of explanation, of an analysis of the modal operator in terms of a quantifier over worlds. But in order genuinely to understand the truth conditions [...] we need to understand what worlds and haecceities are". In our case, the extra syntax in labeled calculi might just create the illusion of providing the process underlying our modal reasoning, by introducing new objects in terms of which to analyze modalities — but they might as well end up distracting from what is really going on. Another issue with this is that possible world semantics not always seems to correspond to why we think modal statements are true. "Given certain purposes, possible worlds semantics may not be the best way to link modal logic to modal reality. [. . .] Why are horses necessarily mammals? Not because every horse is a mammal in every possible world. But because the property of being a horse bears a special relationship to the property of being a mammal" (Warmke, 2016). This might suggest, instead, that Kripke semantics itself has a problem, if it is to provide the semantics of (a fragment of) natural language. But this transfers the burden of the use of semantic elements to semantics itself, which is a different issue. Finally, even if the labeled calculus did reflect a more subconscious aspect of modal reasoning, it is unclear whether this is what Formalization is supposed to achieve. It seems at least a much more dubious way of achieving the goal, as we cannot verify its achievement anymore by looking at natural language itself.

Hence, these matters need attention for the argument that labeled calculi can achieve Formalization. We should take seriously the suggestion that semantic influence in a proof system becomes pollution, precisely when it obstructs the goal of reflecting explicit natural language reasoning.

3.3 Previous analyses

Our general aim is now to find a systematic property of proof systems so that, if their goal is Formalization, the exhibition of this property guarantees that they are semantically polluted. That is, we are aiming to identify when a proof system is too connected to those formal aspects of a model theory that do not have anything to do anymore with informal reasoning.

First, we consider the (few) previous characterizations of semantic pollution in the literature. There are two main definitions of syntactic purity, which is taken as the counterpart of semantic pollution. *Strong syntactic purity* is the satisfaction of Avron (1996)'s previously mentioned requirement, which says that a proof calculus should be independent from a semantics, and that we should not be able to guess the intended semantics just by the form of the structures used in the proof system. Poggiolesi (2010) has named this 'strong' syntactic purity, after she concludes that already the classical propositional sequent calculus is semantically polluted according to this idea. *Weak syntactic purity* is an alternative proposal by Poggiolesi (2010). It says that "[a] sequent calculus should not make use of explicit semantic elements". The idea here is that more proof systems should be able to satisfy this requirement, leading to a more acceptable characterization of semantic pollution. Within this definition, it is the particular understanding of 'explicit semantic element' that will determine the behaviour of semantic pollution among proof systems.

Poggiolesi (2010) proposes that "a sequent (or a set of sequents) does not contain a semantic element if every element that serves to define the sequent (or set of sequents) may be translated in such a way that it forms, together with the translation of the other elements, a formula equivalent to the sequent". That is, a proof system is not semantically polluted if all syntactic elements are reducible back to a previously determined object language (in other words, if it is an *internal* calculus, so that each formula in the proof can be read as a formula of the logic, as opposed to an *external* calculus).

This is a clear characterization, but it is not as explanatory as we would like. It says something mostly about the power of an object language, and classifies everything as either belonging to that language (and being syntactically pure), or as being excluded by it (and being semantically polluted). But take the symbol • ('bullet') in display calculi (originally developed by Belnap (1982), but the bullet was introduced by Wansing (1994)). Depending on its polarity in a sequent, this symbol has the meaning of standard necessity or that of a 'past' diamond. The latter meaning is untranslatable back to the modal object language, but whether it is more of a semantic element than $\Box$ or $\Diamond$ themselves (instead of just something that happens to be unexpressible in the modal language), is not clear. It would be worthwhile to find properties that are more explanatory for what it is that makes a proof system semantically polluted.

4 Candidate properties of semantically polluted calculi

In this section, then, we consider several formal properties of proof systems that have intuitive promise to characterize semantic pollution. We check whether these properties are prevalent among calculi that we think are semantically polluted, such as **G3K**, and scarce among the regular propositional and first-order systems. We also discuss whether certain goals of proof systems are indeed harder to achieve because of these properties. The first two properties we discuss are categoricity of proof systems and proof rules having a semantic translation — they can both be considered as interpretations of 'guessing' a semantics, and so as violations of Avron's strong syntactic purity. The results of these properties are largely negative and show that we need to look in other directions to find the type of characterization we are looking for. The third property concerns a specific type of semantic influence seen in semantic tableaux.

4.1 Categoricity of proof systems

The consideration of categoricity of proof systems is motivated by a strict interpretation of 'strong syntactic purity'. We can imagine that "guessing" the intended semantics of a proof system "from the form of the structures which are used" can occur with different levels of certainty. In case of a low level of certainty, the form of the syntactic structures might just induce some idea or impression of similarity to a semantics — but perhaps a high level of certainty implies that the form of the syntactic structures simply fixes the semantics. The latter idea can be captured by categoricity of proof systems.

Categoricity describes the situation where proof rules determine the truth conditions of their connectives. That is, a proof system is categorical when it is only sound and complete with (uniquely specifies) one collection of admissible valuations. Carnap (1943) describes a well-known counterexample, namely the usual proof system for classical propositional logic. This proof system is perfectly consistent with the non-standard valuation that makes everything true (in particular, for each formula A, both A and $\neg A$ will come out true, going against the desired truth conditions for negation).

To weaken the property of categoricity a bit, we will also talk about categoricity of a *connective*. A connective is categorical when a proof system is categorical — or when a proof system is not categorical, yet non-standard valuations for the particular connective in question are excluded. Thus, the inference rules for a particular connective do determine the semantics of this connective, even though the proof system as a whole is not categorical. This suggests we could take two variations on a requirement for semantic pollution.

Definition 1 (Semantic pollution in terms of categoricity) *A proof system is* strongly *semantically polluted if it is categorical. A proof system is* weakly *semantically polluted if it is not categorical, but it does have categorical connectives.*

However, it is quickly seen that this definition of semantic pollution is not easily satisfied by proof systems. Few proof systems are categorical, unless they are specifically designed to be so by acting on the proof system, the semantics, or on the relationship between proof systems and semantics. Making sure a proof system is multiple-conclusion is one way to ensure categoricity (Rumfitt, 1997); other ways are adding a primitive notion of rejection (Smiley, 1996), or working with n-sided sequents (Hjortland, 2014). But categoricity can also be ensured by restricting the interpretation space on the semantic side, by imposing certain semantic principles (Bonnay & Westerståhl, 2016). These, however, all seem artificial methods that have very little to do with semantic notions entering a proof system. That is, the strengthening of proof systems to multiple-conclusion or bilateralist calculi does not involve any referral to semantic notions (such as possible worlds or the Kripke accessibility relation). For multiple-conclusion calculi, the reason that they become categorical seems almost accidental: it is the difference between the empty conclusion being vacuously true for single-conclusion calculi, while it becomes false for multiple conclusions (as truth demands that *some* conclusion must be true, see also Rumfitt, 1997). This seems a technical circumstance that does not seem to hinge on the introduction of multiple conclusions themselves, but more on the choice of their meta-theoretical interpretation. The literature furthermore shows that the standard calculi for propositional, modal and first-order logic are only weakly categorical, except somehow intuitionistic propositional systems, which are strongly categorical (Tong & Westerståhl, 2023).

Additionally, categoricity is generally treated as a desirable property, as we want proof calculi to capture the meaning of their connectives and to exclude unwanted interpretations. This is in line with what we see as an indirect but positive relation between categoricity and the ability of a proof system to satisfy Formalization. If a proof system rules out non-standard valuations, this strengthens the idea that it can be a system open to human understanding and use. It is at least tempting to think that the way we reason in practice is more formed by entities that have a consistent inherent meaning, than by entities that may have 'arbitrarily' managed to slip into our vocabulary. Thus, a categoricity result seems far from likely to identify elements in the proof system that are not used in informal reasoning.

4.2 Semantic translation of proof rules

The second property of proof systems we consider is the existence of a way to 'read off' the semantics from the inference rules. A looser interpretation of 'guessing' the semantics from a proof system gives rise to this. Given a certain familiarity with the intended semantics of a proof system, the truth conditions of a connective may be recognizable in the proof rules. Such a connection between the proof system and a semantics, instead, might be reason to call a proof system semantically polluted. Poggiolesi (2010) discusses this idea:

> (Poggiolesi, 2010) [...] the logical rules of $\mathbf{Gcl}_L$ *reflect* at the syntactic level (or may be read in terms of) the semantic definitions of each constant: the elements of the structure of the sequent calculus (i.e. the sequent arrow and the comma) remind us of the metalinguistic elements of the definitions (i.e. *if ... then* and *and* and *or*); the positions of the formulas in the sequent (i.e. the left or the right sides of the sequent) remind us of the truth values in the equivalencies (i.e. false or true).

This can be made more precise for classical propositional logic by Hacking (1979)'s Do-It-Yourself semantics. Hacking assumes a formal language and a model theory for that language, and presents a way to go from general proof rules to the truth conditions of the connectives defined. His main example

Robin Martinot

concerns sequent rules (omitting side formulas) for a new connective added to classical propositional logic, which occurs in the principal formula φ:

$$\frac{\{\Gamma_i \Rightarrow \Delta_i\}_{i \in I}}{\Rightarrow \varphi} \qquad \frac{\{\Gamma_j \Rightarrow \Delta_j\}_{j \in J}}{\varphi \Rightarrow}$$

Given such rules, we can tell when φ is true or false in a model of the original language: the left rule tells us that φ is true in a model of the original language iff there are premises $\{\Gamma_i \Rightarrow \Delta_i\}_{i \in I}$ from which to derive $\Rightarrow \varphi$, and if for each premise $\Gamma_i \Rightarrow \Delta_i$ it holds that either some $\gamma \in \Gamma_i$ is false, or some $\delta \in \Delta_i$ is true. The other rule similarly tells us when φ is false in a model of the previous language. This inspires the following more general notion of semantic pollution.

Definition 2 (Semantic pollution in terms of translations) *A proof system is semantically polluted if, given a model theory for its formal language, there is a systematic translation[2] that takes the inference rules for each (existing or new) logical connective to the truth condition of this connective.*

For our purposes, this property is too easily satisfied by proof systems, and simultaneously not developed enough for wide applications. Hacking relies on strong assumptions for the existence of such a translation method. He presupposes that cut-free derivations are available, and that rules are 'local' in the sense that "they concern only the components from which the principal formula is built up, and place no restrictions on the side formulas" (Hacking, 1979). Hacking's resort to the ω-rule for first-order logic then turns out to enforce classicality (Sundholm, 1981). In short, the "DIY semantics" in practice comes down to quite a restricted notion, and applicable almost only to certain classical propositional systems. Perhaps more importantly, and already noted by Poggiolesi (2010), is that this property tells us that proof systems for classical propositional logic are semantically polluted. This is reason enough to discard the property as a serious candidate for corresponding to semantic pollution as we think of it intuitively.

The property seems to have an inconclusive relation to Formalization. Having a semantic translation shows a clear characterization of when a connective is true and false. But this transfers the decision of pollution to semantics itself: if a truth condition contains many purely model-theoretic

[2]There are different ways of making this translation precise: clearly, some preservation of structure is required. However, we will not elaborate on the properties of such a translation, as Hacking's example will already provide reason to reject the measure.

notions, an inference rule from which to read off this semantics must some-how represent these ingredients. Instead, if a truth condition is closely tied to natural language, having a semantic translation should only be helpful to Formalization. This reinforces the idea that semantic translation of proof rules is also not a property indicating semantic pollution.

4.3   Meta-theoretic pollution

We finally mention some properties of proof that are less systematic, but may deserve further attention in relation to a more meta-theoretical type of semantic pollution. It has to do with what we think a formal proof is allowed to be: before, we took it as a given that a proof system had a good sense of what an axiom is, a conclusion and an inference rule, and that this did not affect semantic pollution. Here, we want to consider the possibility that these choices can cause a form of semantic pollution.

First, most proof systems take the axioms of a proof to represent the ba-sic principles we have of a logic or theory when we start any deduction. Additional assumptions may be made in case we would like to analyze their specific consequences. Different proof systems (such as Hilbert, sequent and natural deduction calculi) incorporate the axioms and assumptions in differ-ent ways – however, the general pattern is that a proof of a theorem A works from tautological axioms and possibly assumptions towards the theorem by applying inference rules to the axioms. By soundness and completeness, such a proof tells us that on the model-theoretic side, each model satisfies A. Thus, it seems that a formal proof normally has a general (universal) connection to the model theory.

However, in order to use proof systems such as semantic tableaux (see, e.g. Priest, 2008), axioms include not only possible assumptions, but also the negation of the conclusion. And instead of working towards the theorem, the proof works towards a contradiction. This can be seen as a different perspective on what the axioms and conclusion of a proof should be. It turns out that semantic tableaux can have a more *specific* connection to the model theory, by providing a method to identify a specific (counter)model from a formal proof. This may indicate a type of semantic pollution, in the sense that it shows a more precise connection to the model theory.

Second, we comment on the form of inference rules that are allowed to be part of a proof system. For example, as mentioned, Steinberger (2011) argues that multiple-conclusion rules do not reflect our inferential practices. Consider also the ω-rule.

$$\frac{\varphi(0) \qquad \varphi(1) \qquad \varphi(2) \qquad \ldots}{\forall n\varphi(n)}$$

Although most people seem to think that the ω-rule is intuitive, and "instances of the omega-rule are at least informally valid" (McGee, cited in Murzi, 2014), it is also thought to possess semantic content. This is because the rule works from infinitely many premises, and "it stretches the concept of a rule to encompass rules nobody is every able to apply" (Peregrin, 2020). Interestingly, this is a way in which this rule can be seen as violating Formalization. Only the semantic side seems fully equipped to deal with full infinity, where models may well be infinite mathematical structures. Thus, the infinitary aspect of semantics could be a property that can pollute a proof system. It would be interesting to see whether other properties of models can invade a proof theory in this way.

In brief, differences in the form of axioms, inference rules and conclusions of a proof may induce a closer connection to models and semantic notions, and characterize different types of semantic pollution.

5   Conclusion

In this paper, we emphasize the importance of knowing the goal of a proof system when designing it. This seems increasingly important for the recent development of many generalizations of proof calculi, so that it remains clear what they should and should not be used for. Properties of the proof system can then be evaluated by how well they facilitate achieving that goal. We propose that goals of proof systems can be categorized under Provability, Applicability, Formalization, Normativity and Inferentialism.

We additionally suggest that semantic pollution can be seen as a proof system that, in different guises, cannot fully achieve Formalization, because of the incorporation of model-theoretic elements that do not belong to informal reasoning. We discuss various candidate properties that might more systematically characterize semantic pollution, but conclude that neither categoricity of a proof system, nor having a semantic translation of proof rules suffice as a measure. Finally, we suggest that semantic tableaux and the ω-rule might give reason to introduce a different, more meta-theoretical, type of semantic pollution. Future research should tell us whether there is a better and more explanatory analysis of the semantic pollution we see in labeled systems. It remains a challenge to characterize semantic pollution by properties that are clear and reliable, yet that also genuinely concern semantic notions, instead of exposing an irrelevant correlation to semantics.

References

Arana, A. (2009). On formally measuring and eliminating extraneous notions in proofs. *Philosophia Mathematica*, *17*(2), 189–207.

Avron, A. (1996). The method of hypersequents in the proof theory of propositional non-classical logics. In W. Hodges, M. Hyland, C. Steinhorn, & J. Truss (Eds.), *Logic: From Foundations to Applications* (pp. 1–32). Oxford: Oxford University Press.

Belnap, N. D. (1982). Display logic. *Journal of Philosophical Logic*, *11*, 375–417.

Bimbó, K. (2014). *Proof Theory: Sequent Calculi and Related Formalisms*. Boca Raton: CRC Press.

Bonnay, D., & Westerståhl, D. (2016). Compositionality solves Carnap's problem. *Erkenntnis*, *81*(4), 721–739.

Buss, S. R. (1998). An introduction to proof theory. In S. R. Buss (Ed.), *Handbook of Proof Theory* (Vol. 137, pp. 1–78). Amsterdam: Elsevier.

Carnap, R. (1943). *Formalization of Logic*. Cambridge, MA: Harvard University Press.

Fitting, M., & Kuznets, R. (2015). Modal interpolation via nested sequents. *Annals of Pure and Applied Logic*, *166*(3), 274–305.

Hacking, I. (1979). What is logic? *The Journal of Philosophy*, *76*(6), 285–319.

Hjortland, O. T. (2014). Speech acts, categoricity, and the meanings of logical connectives. *Notre Dame Journal of Formal Logic*, *55*(4), 445–467.

Maddy, P. (2019). What do we want a foundation to do? In S. Centrone, D. Kant, & D. Sarikaya (Eds.), *Reflections on the Foundations of Mathematics* (pp. 293–311). Cham: Springer.

Menzel, C. (1990). Actualism, ontological commitment, and possible world semantics. *Synthese*, *85*, 355–389.

Murzi, J. (2014). The inexpressibility of validity. *Analysis*, *74*(1), 65–81.

Negri, S. (2005). Proof analysis in modal logic. *Journal of Philosophical Logic*, *34*(5), 507–544.

Negri, S. (2011). Proof theory for modal logic. *Philosophy Compass*, *6*(8), 523–538.

Negri, S. (2016). Non-normal modal logics: A challenge to proof theory. In P. Arazim & T. Lávička (Eds.), *The Logica Yearbook 2016* (pp. 125–140). London: College Publications.

Negri, S. (2017). Proof theory for non-normal modal logics: The neighbourhood formalism and basic results. *IfCoLog Journal of Logics and their Applications*, *4*(4), 1241–1286.

Peregrin, J. (2012). What is inferentialism. In L. Gurova (Ed.), *Inference, Consequence, and Meaning: Perspectives on Inferentialism* (pp. 3–16). Newcastle upon Tyne: Cambridge Scholars Publishing.

Peregrin, J. (2020). Rudolf Carnap's inferentialism. In R. Schuster (Ed.), *The Vienna Circle in Czechoslovakia* (pp. 97–109). Cham: Springer.

Poggiolesi, F. (2009). The method of tree-hypersequents for modal propositional logic. In D. Makinson, J. Malinowski, & H. Wansing (Eds.), *Towards Mathematical Philosophy* (Vol. 28, pp. 31–51). Dordrecht: Springer.

Poggiolesi, F. (2010). *Gentzen Calculi for Modal Propositional Logic* (Vol. 32). Dordrecht: Springer.

Poggiolesi, F. (2020). A proof-based framework for several types of grounding. *Logique et Analyse*, *252*, 387–414.

Priest, G. (2008). *An Introduction to Non-Classical Logic: From If to Is*. Cambridge: Cambridge University Press.

Read, S. (2015). Semantic pollution and syntactic purity. *The Review of Symbolic Logic*, *8*(4), 649–661.

Rumfitt, I. (1997). The categoricity problem and truth-value gaps. *Analysis*, *57*(4), 223–235.

Smiley, T. (1996). Rejection. *Analysis*, *56*(1), 1–9.

Steinberger, F. (2011). Why conclusions should remain single. *Journal of Philosophical Logic*, *40*(3), 333–355.

Sundholm, G. (1981). Hacking's logic. *The Journal of Philosophy*, *78*(3), 160–168.

Tong, H., & Westerståhl, D. (2023). Carnap's problem for intuitionistic propositional logic. *Logics*, *1*(4), 163–181.

Tosatto, S. C., Boella, G., van der Torre, L., & Villata, S. (2012). Abstract normative systems: Semantics and proof theory. In *Principles of Knowledge Representation and Reasoning: Proceedings of the Thirteenth International Conference (KR 2012)* (pp. 358–368).

van der Torre, L., & Villata, S. (2014). An ASPIC-based legal argumentation framework for deontic reasoning. In S. Parsons, N. Oren, C. Reed, & F. Cerutti (Eds.), *Computational Models of Argument: Proceedings of COMMA 2014* (pp. 266–421). Amsterdam: IOS Press.

Wansing, H. (1994). Sequent calculi for normal modal propositional logics. *Journal of Logic and Computation, 4*(2), 125–142.

Warmke, C. (2016). Modal semantics without worlds. *Philosophy Compass, 11*(11), 702–715.

Robin Martinot
Utrecht University, Department of Theoretical Philosophy and Religious Sciences
The Netherlands
E-mail: r.a.martinot@uu.nl

Symmetry Principles in Pure Inductive Logic

JEFF PARIS AND ALENA VENCOVSKÁ

Abstract: This article is a contribution to the founding problem of pure inductive logic: what probability function on the set of sentences of a predicate language L should be chosen by a rational agent who knows nothing about any intended interpretation of the non-logical symbols of the language. In particular, we consider some symmetry principles proposed for such a prior probability function to date. These include the 'ultimate' symmetry principle INV which, although seemingly very well justified, proved too strong to be useful, and also its pragmatic restriction PIP, equivalent to the principle Ax for unary languages. We analyse what is unreasonable in the formulation of INV and we arrive at a modification of it, ENV. We show that in the case of unary languages, ENV is equivalent to Ax but in the polyadic context, ENV is strictly stronger than PIP.

Keywords: Pure Inductive Logic, Symmetry Principles, Exchangeable Invariance Principle, Permutation Invariance Principle, Atom Exchangeability Principle, Rudolf Carnap

1 Introduction

This paper concerns pure inductive logic, an area that has developed from the foundations laid by Rudolf Carnap during the middle third of the twentieth century. Carnap's ambitious programme aimed to capture all inductive reasoning about the world in terms of inductive logic, just as deductive reasoning has been captured by predicate logic. It involved a move from the world and its properties that we perceive, about which we wish to reason and to which we attach probabilities, to a formal system that we can investigate analytically. We shall leave aside the considerable difficulties and problems encountered in the development of Carnap's programme over the course of time but we note that Carnap was quite consistent in promoting pure inductive logic, that is, a purely logical and mathematical investigation of the theory of logical probability as an essential part of his programme.

99

Jeff Paris and Alena Vencovská

This involves the study of properties of prior probability functions on the set of sentences of a formal language that satisfy certain formal principles. The principles can be motivated by considering the imagined behaviour of an ideally rational agent in a situation of total ignorance. Interest in such probability functions stems from the idea that a rational agent only needs to be provided with an initial rational prior probability distribution and subsequently they can accommodate all incoming information via successive conditioning on the probability function currently held.

In recent times, pure inductive logic has undergone rapid development, emboldened and liberated by its obvious relevance to artificial intelligence. While Carnap concentrated his efforts on languages with unary predicates only (remarkable success in the area being the derivation of what came to be called Carnap's Continuum of Inductive Methods, see Johnson, 1932 or Carnap & Stegmüller, 1959), modern pure inductive logic has included polyadic languages. Numerous principles for the desired prior probability function to satisfy have been proposed and investigated, falling roughly under three headings, and referred to as principles of symmetry, relevance and irrelevance. This paper concentrates on the principles of symmetry. [1]

Beyond individual instances of symmetry as regards the constant symbols, relation symbols, relation symbols versus their negations, and, in the unary case, atoms (conjunctions of literals), an 'ultimate' symmetry principle has been proposed that came to be called the Invariance Principle. It is based on the argument that any valid symmetry stems from some automorphism of the underlying mathematical framework, and that conversely, any such automorphism represents a symmetry that should be respected. The principle, albeit its effects are poorly understood in the case of polyadic languages, has, in a sense, turned out to be too strong, and for some time it appeared that it would be necessary to replace it by a restricted version of itself, the Permutation Invariance Principle. Although remarkably well recommended by other independent considerations, the Permutation Invariance Principle appears to lack the appeal of inevitability that the Invariance Principle carries. Recently, an argument has been proposed throwing a new light on why the Invariance Principle is too strong and how the original formulation should be amended. It has lead to a new candidate for the 'ultimate' symmetry principle, the Exchangeable Invariance Principle. In this contribution, we survey

---

[1] The terminology is a little confusing: Carnap's Axiom of Symmetry is the modern Principle of Constant Exchangeability, Ex, see the next section. Carnap refers to what are now symmetry principles as *principles of invariance* (see e.g. Carnap, 1962), noting that they are what should capture 'the valid core of the old principle of indifference (or principle of insufficient reason)'.

these matters and show that in the unary case, the Exchangeable Invariance Principle is in fact equivalent to the Permutation Invariance Principle (and thus also simply to the requirement of symmetry with respect to atoms, the Atom Exchangeability Principle). We also show that in the polyadic case, the Exchangeable Invariance Principle is strictly stronger than the Permutation Invariance Principle.

2 Notation and background

We work with a first order language L with relation symbols $R_1, \ldots, R_q$ of arities $r_1, \ldots, r_q$ respectively, constants a_n for $n \in \mathbb{N}^+ = \{1, 2, 3, \ldots\}$, and no function symbols nor the equality symbol. We write $L = \{R_1, \ldots, R_q\}$ for such a language. To avoid double subscripts we sometimes use $b_1, b_2, \ldots, b_m$ etc. in place of $a_{i_1}, a_{i_2}, \ldots, a_{i_m}$, and in the special case of a language that contains unary predicates only (no relation symbols of higher arities) we sometimes use $P_1, \ldots, P_q$ rather than $R_1, \ldots, R_q$ to emphasise the unary context. The set of sentences L is denoted SL, and its elements are denoted θ, ϕ, ψ etc. Writing, for example, $\psi(b_1, \ldots, b_n)$ usually means that all the constants appearing in ψ are amongst $b_1, \ldots, b_n$, unless indicated otherwise.

A function $w : SL \to [0, 1]$ is a probability function on SL if for all $\theta, \phi, \exists x\, \psi(x) \in SL$,

(P1) $\vDash \theta \implies w(\theta) = 1$.

(P2) $\vDash \neg(\theta \wedge \phi) \implies w(\theta \vee \phi) = w(\theta) + w(\phi)$.

(P3)[2] $w\left(\exists x\, \psi(x)\right) = \lim_{n \to \infty} w(\bigvee_{i=1}^{n} \psi(a_i))$.

An important point to note is that probability functions respect logical equivalence, that is, logically equivalent sentences get the same probability.

We are interested in symmetry principles that have been proposed as rational requirements for a prior probability function of an ideally logical or rational agent who has no information about any intended interpretation of the language.

The most fundamental of these principles is that of *Constant Exchangeability* (Carnap's Axiom of Symmetry), which may be formulated as follows:

---

[2] There may be some constants appearing in $\psi(x)$, so in this condition, writing $\psi(a_i)$ is an exception to the convention stated just above.

Jeff Paris and Alena Vencovská

Principle of Constant Exchangeability, Ex. If g is a permutation of $\mathbb{N}^+$ then

$$w(\theta(a_1, a_2, \ldots, a_m)) = w(\theta(a_{g(1)}, a_{g(2)}, \ldots, a_{g(m)})).$$

Other basic symmetry principles include

Predicate Exchangeability Principle, Px. If R, R' are relation symbols of L of the same arity then for $\theta \in SL$,

$$w(\theta) = w(\theta')$$

where θ' is the result of simultaneously replacing R by R' and R' by R throughout θ.

Strong Negation Principle, SN. For $\theta \in SL$,

$$w(\theta) = w(\theta')$$

where θ' is the result of replacing each occurrence of R in θ by $\neg R$.

For the special case of $L = \{P_1, \ldots, P_q\}$ with only unary predicates there is also the *Principle of Atom Exchangeability*, formulated in terms of unary atoms:

An atom of $L = \{P_1, \ldots, P_q\}$ is a formula of the form

$$\bigwedge_{i=1}^{q} \pm P_i(x)$$

where $\pm P$ stands for one of P or $\neg P$. They are denoted $\alpha_1(x), \ldots, \alpha_{2^q}(x)$.

Atom Exchangeability Principle, Ax. For L and α_i as above, if h is a permutation of $\{1, 2, \ldots, 2^q\}$ and $j_1, \ldots, j_m \in \{1, 2, \ldots, 2^q\}$ then

$$w(\alpha_{j_1}(b_1) \wedge \ldots \wedge \alpha_{j_m}(b_m)) = w(\alpha_{h(j_1)}(b_1) \wedge \ldots \wedge \alpha_{h(j_m)}(b_m)).$$

3 Framework automorphisms and invariance principles

In this section, we shall survey some results that can be found in (Paris & Vencovská, 2015) and (Paris & Vencovská, 2019). Motivation for these arguments came from the following considerations: Imagine an ideally logical agent who knows that they are in a structure M for L. The agent does not know which structure they are in but is 'aware' of all the possibilities. They need to assign, in a rational way, probabilities to all sentences of the language expressing their belief in each sentence holding in their particular structure.

This amounts to asking: For a given language L, what probability function on SL should a logical agent in this situation choose?

Let $\mathcal{T}$ denote the set of structures for L with universe $\{a_1, a_2, a_3, \ldots\}$, with each a_i interpreted as itself. These structures capture all in L that can be true in the sense that if $\psi \in SL$ is consistent then there is $M \in \mathcal{T}$ such that $M \models \psi$.[3] Hence we have

$$\vdash \psi \iff \models \psi \iff \forall M \in \mathcal{T},\ M \models \psi.$$

For $\phi \in SL$ let
$$[\phi] = \{M \in \mathcal{T} : M \models \phi\}.$$

Note that logically equivalent sentences define the same subset of $\mathcal{T}$ and for logically valid sentences ϕ, $[\phi]$ is the whole of $\mathcal{T}$ and for contradictions it is the empty set. Also,

$$[\phi \wedge \theta] = [\phi] \cap [\theta], \quad [\phi \vee \theta] = [\phi] \cup [\theta], \quad [\neg\phi] = \mathcal{T} \setminus [\phi].$$

The collection of all $[\phi]$ for $\phi \in SL$ forms an algebra of subsets of $\mathcal{T}$, denoted $\mathcal{B}$. We define BL to be the 2-sorted structure consisting of $\mathcal{T}$ and its subsets from $\mathcal{B}$.

Thus what the agent is 'aware' of is the 2-sorted grand structure BL. Asked to assign probabilities to sentences, the agent could check from this picture for example that a logically valid sentence ϕ must get value 1 because whichever structure M the agent might be in, the sentence holds in it. Similarly for contradictions. With logically contingent sentences the exact values cannot be derived from this picture alone, but the symmetries in it dictate that a rational agent should assign certain pairs of sentences the same values, namely those which can be mapped onto each other by an automorphism of the above grand structure BL.

An automorphism of BL is a one-one function F from $\mathcal{T}$ onto $\mathcal{T}$ which also permutes $\mathcal{B}$, that is: For each sentence ϕ of L there is a sentence ψ of L

---

[3]This follows since the countability of the language L means that ψ must have a countable model, and hence, by re-interpreting the constant symbols not mentioned in ψ as the remaining a_i, a model in $\mathcal{T}$.

such that

$$M \models \phi \iff F(M) \models \psi, \text{ that is, } \{F(M) : M \models \phi\} = \{M : M \models \psi\}$$

and conversely, for each ψ there is some ϕ such that the above holds.

Somewhat abusing notation, we write $F(\phi)$ for such ψ. This is determined up to logical equivalence. As can easily be checked,

$$\{F(M) : M \models \phi \wedge \theta\} = \{F(M) : M \models \phi\} \cap \{F(M) : M \models \theta\},$$

so $F(\phi \wedge \theta) = F(\phi) \wedge F(\theta)$, and similarly $F(\theta \vee \phi) = F(\theta) \vee F(\phi)$ and $F(\neg\phi) = \neg F(\phi)$ and F is an automorphism of the Lindenbaum algebra of L (but there are automorphisms of the Lindenbaum algebra of L which are not automorphisms of BL, see Paris & Vencovská, 2017).

Since for any such automorphism, a sentence and its image would look the same to an agent with only BL at their disposal, they should attach the same probability to each of them. Stated more formally, this is the *Invariance Principle* of pure inductive logic:

Invariance Principle, INV. If F is an automorphism of BL and $\theta \in SL$ then

$$w(\theta) = w(F(\theta)).$$

INV implies Ex, Px, SN. In the unary context, it implies Ax. But it is even much stronger than that. In (Paris & Vencovská, 2011) the following result has been proved:

For a unary language $L = \{P_1, \ldots, P_q\}$, INV reduces the 'rational choices' to a single probability function, c_0, where

$$c_0 \left(\bigwedge_{k=1}^{m} \alpha_{j_k}(b_k) \right) = \begin{cases} 2^{-q} & \text{if } j_1 = j_2 = \ldots = j_m, \\ 0 & \text{otherwise.} \end{cases}$$

This suggests that the Invariance Principle is too strong. Its power can be reduced by imposing some conditions on the automorphisms of BL that are required to be respected, as indeed one can easily see when endeavouring to derive, for example, Ex from INV. We will now describe one natural possibility of doing this which, coupled with Ex, yields a symmetry principle stronger than those considered previously, and yet not too strong. First we need a few more definitions.

A *state description* for constants $b_1, \ldots, b_n$ is a sentence $\Theta(b_1, \ldots, b_n)$ that decides all atomic sentences involving these constants. In other words, if the language has relation symbols $R_1, R_2, \ldots, R_q$ with arities $r_1, r_2, \ldots, r_q$ respectively, then such a state description has the form

$$\bigwedge_{s=1}^{q} \bigwedge_{i_1,i_2,\ldots,i_{r_s}=1}^{n} \pm R_s(b_{i_1}, b_{i_2}, \ldots, b_{i_{r_s}})$$

where $\pm R$ stands for one of R or $\neg R$.

A *state formula* is as a state description only with variables replacing constants.

A (polyadic) *atom* is a state formula $\gamma(x_1, \ldots, x_r)$ for r variables, where r is the maximal arity of a relation symbol in L.

Note that when L is unary, the definition of an atom agrees with the definition of a unary atom given earlier. We continue to denote them $\alpha_1(x), \ldots, \alpha_{2^q}(x)$. So for unary L, a state description for $b_1, \ldots, b_n$ is (logically equivalent to) a sentence of the form $\alpha_{j_1}(b_1) \wedge \alpha_{j_2}(b_2) \wedge \ldots \wedge \alpha_{j_n}(b_n)$.

An automorphism F of BL *permutes atoms*[4] if there is a permutation f of atoms such that for any atom γ and constants $b_1, \ldots, b_r$,

$$F(\gamma(b_1, \ldots, b_r)) = f(\gamma)(b_1, \ldots, b_r).$$

f determines F but if the language is not purely unary only some permutations of atoms actually do generate automorphisms of BL.

Permutation Invariance Principle, PIP. If F is an automorphism of BL that permutes atoms and $\theta \in SL$ then

$$w(\theta) = w(F(\theta)).$$

Interestingly, PIP has two other equivalent formulations that, entirely informally, can be expressed as follows

• If a sentence is a *translated version* (see Paris & Vencovská, 2019) of another sentence then they should have the same probability (TIP).

• State descriptions *with the same structure*[5] should have the same probability (NIP).

---

[4]The original and equivalent definition involved permuting state formulae, see (Paris & Vencovská, 2015), and (Ronel & Vencovská, 2014) for details regarding the equivalence.

[5]That is, *similar* state descriptions, see (Paris & Vencovská, 2015).

Jeff Paris and Alena Vencovská

4 Exchangeable automorphisms

The Permutation Invariance Principle PIP has much to recommend it as a replacement of INV, but in common with other symmetry principles it lacks the status of being the ultimate symmetry principle, the one to stop at and declare to be the explication of symmetry within pure inductive logic. For why stop at automorphisms permuting atoms, even if the principle obtained in this way has multiple other independent justifications?

With the original motivation for INV in mind we consider the way in which it was shown (at least in the unary case) that almost no probability functions satisfy it. We note that it involved automorphisms defined via singling out some constant symbols and letting the corresponding constants play a special role - across structures, beyond the roles that they have in the individual structures. However, since the agent is not supposed to give a special status to any particular constants, it appears that they should not be 'aware' of such automorphisms. This leads us to define a restricted class of automorphisms as follows:

An automorphism F of BL is *exchangeable* if whenever δ is a permutation of $\mathbb{N}^+$, $\theta(a_1, \ldots, a_n) \in SL$, and

$$F(\theta(a_1, a_2, \ldots, a_n)) = \phi(a_{i_1}, a_{i_2}, \ldots, a_{i_m})$$

then

$$F(\theta(a_{\delta(1)}, a_{\delta(2)}, \ldots, a_{\delta(n)})) = \phi(a_{\delta(i_1)}, a_{\delta(i_2)}, \ldots, a_{\delta(i_m)}).$$

The following lemma shows an important property of exchangeable automorphisms.

Lemma 1 *If $\psi = \psi(b_1, b_2, \ldots, b_m) \in SL$ then $F(\psi)$ can be taken to mention only constants from amongst $b_1, b_2, \ldots, b_m$ (since in the equivalence class of sentences referred to as $F(\psi)$ there are sentences mentioning only constants from amongst $b_1, b_2, \ldots, b_m$).*

Proof. Suppose that $F(\psi(\vec{b})) = \phi(\vec{b}, \vec{c})$ where $\vec{c}$ are some constants not mentioned amongst the $\vec{b} = b_1, \ldots, b_m$. Then by considering permutations which fix the $\vec{b}$, for any other new constants $\vec{d}$, we would have

$$\vDash \phi(\vec{b}, \vec{c}) \leftrightarrow \phi(\vec{b}, \vec{d}).$$

Hence

$$\vdash \phi(\vec{b}, \vec{c}) \leftrightarrow \phi(\vec{b}, \vec{d})$$

and by replacing $\vec{d}$ in this formal proof by the free variables $\vec{x}$

$$\vdash \phi(\vec{b}, \vec{c}) \longleftrightarrow \phi(\vec{b}, \vec{x}).$$

So

$$\vdash \phi(\vec{b}, \vec{c}) \longleftrightarrow \forall \vec{x}\, \phi(\vec{b}, \vec{x})$$

and

$$\models \phi(\vec{b}, \vec{c}) \longleftrightarrow \forall \vec{x}\, \phi(\vec{b}, \vec{x}),$$

giving

$$F(\psi) = \phi(\vec{b}, \vec{c}) = \forall \vec{x}\, \phi(\vec{b}, \vec{x}).$$

$\square$

Exchangeable automorphisms aim to capture symmetries of the framework beyond the symmetry in constants. The symmetry in constants is carried by arbitrary permutations of constants. Thus it now appears that the ultimate symmetry principle duo should be Ex along with the principle defined below.

Exchangeable Invariance Principle, ENV. If F is an exchangeable automorphism of BL and $\theta \in SL$ then

$$w(\theta) = w(F(\theta)).$$

Note that ENV implies Px and SN and PIP (and hence, in the unary case, Ax). This follows in the same way as the corresponding results for INV, see (Paris & Vencovská, 2015), since the automorphisms to which we need to restrict INV to obtain these principles, are exchangeable.

5 The Exchangeable Invariance Principle for unary languages

In this section, we shall assume that L is unary, with atoms $\alpha_i(x)$ for $1 \le i \le 2^q$. For $S \subseteq \{1, \ldots, 2^q\}$, $S \neq \emptyset$, let $\overline{S}$ stand for $\{1, \ldots, 2^q\} \setminus S$ and let

$$\exists S \neg \exists \overline{S}$$

stand for

$$\bigwedge_{i \in S} \exists x\, \alpha_i(x) \wedge \neg \exists x \bigvee_{i \in \overline{S}} \alpha_i(x).$$

The following result is well known but for a detailed proof see for example (Paris, 2016).

Jeff Paris and Alena Vencovská

Lemma 2 *If $\psi \in SL$ and the constants mentioned in ψ are amongst $b_1, b_2, \ldots, b_m$ then ψ is logically equivalent to a sentence of the form*

$$\bigvee_{k=1}^{h} \left(\bigwedge_{i=1}^{m} \alpha_{f_{ki}}(b_i) \wedge \exists S_k \neg \exists \overline{S_k} \right) \tag{1}$$

where all these disjuncts are disjoint and satisfiable.

Now assume that F is an exchangeable automorphism. From Lemmas 1 and 2, for $\emptyset \neq S \subseteq \{1, \ldots, 2^q\}$

$$F(\exists S \neg \exists \overline{S}) = \bigvee_{k=1}^{h} \exists S_k \neg \exists \overline{S_k} \quad \text{for some } S_1, S_2, \ldots, S_h.$$

So since the left hand sides are disjoint for different S, by the Pigeonhole Principle we must have

$$F(\exists S \neg \exists \overline{S}) = \exists V \neg \exists \overline{V} \quad \text{for a single } V.$$

Carrying on with this notation, again by Lemmas 1 and 2 and the above, for $i \in S$ there is a set $\emptyset \neq K_i \subseteq V$ such that

$$F(\alpha_i(a_1) \wedge \exists S \neg \exists \overline{S}) = \bigvee_{j \in K_i} (\alpha_j(a_1) \wedge \exists V \neg \exists \overline{V})$$

(the right hand side must feature a_1 since otherwise it would be just $\exists V \neg \exists \overline{V}$ which is not possible) so since the left-hand sides are disjoint for different i, we have $|S| \leq |V|$. Thus, continuing to use pigeonhole arguments, for $|S| = 2^q$ we have $S = V$ and for $|S| = 2^q - 1$ we have $|S| = |V|$ and thus $V = \tau_{2^q-1}(S)$ for some permutation τ_{2^q-1} of the subsets of $\{1, \ldots, 2^q\}$ with $2^q - 1$ elements. Also, each K_i is a singleton and

$$F(\alpha_i(a_1) \wedge \exists S \neg \exists \overline{S}) = \alpha_{\sigma_S(i)}(a_1) \wedge \exists V \neg \exists \overline{V}$$

for some bijective mapping σ_S from S onto V. Continuing down with $|S|$, we see that there is a permutation τ of the powerset of $\{1, \ldots, 2^q\}$ preserving cardinality (that is, $|\tau(S)| = |S|$ for each S), and bijections σ_S from S to $\tau(S)$ such that for any S and $i \in S$,

$$F(\alpha_i(a_1) \wedge \exists S \neg \exists \overline{S}) = \alpha_{\sigma_S(i)}(a_1) \wedge \exists \tau(S) \neg \exists \overline{\tau(S)}.$$

Symmetry Principles in Pure Inductive Logic

F determines τ and the σ_S, and, again using Lemma 2, arbitrary τ and σ_S determine F since the unique way of extending it is via sending (1) to

$$\bigvee_{k=1}^{h}\left(\bigwedge_{i=1}^{m}\alpha_{\sigma_{S_k}(f_{ki})}(b_i) \wedge \exists\tau(S_k)\neg\exists\overline{\tau(S_k)}\right).$$

Now assume that w satisfies Ax. We shall show that this suffices to guarantee that for $S, V \subseteq \{1, \ldots, 2^q\}$, a bijection σ from S onto V and $f_1, f_2, \ldots, f_m \in S$,

$$w\left(\bigwedge_{i=1}^{m}\alpha_{f_i}(b_i) \wedge \exists S\neg\exists\overline{S}\right) = w\left(\bigwedge_{i=1}^{m}\alpha_{\sigma(f_i)}(b_i) \wedge \exists V\neg\exists\overline{V}\right). \qquad (2)$$

By (P3), for $k \in \{1, 2, \ldots, 2^q\}$,

$$w(\exists x\, \alpha_k(x)) = \lim_{n\to\infty} w\left(\bigvee_{i=1}^{n}\alpha_k(a_i)\right) = \lim_{n\to\infty} w\left(\bigvee T_n^k\right)$$

where T_n^k consists of all conjunctions

$$\alpha_{r_1}(a_1) \wedge \alpha_{r_2}(a_2) \wedge \ldots \wedge \alpha_{r_n}(a_n)$$

where at least one of the $r_j = k$. Notice that since $\vDash \bigvee T_n^k \to \exists x\, \alpha_k(x)$, this gives that

$$\lim_{n\to\infty} w\left(\exists x\, \alpha_k(x) \leftrightarrow \bigvee T_n^k\right) = 1.$$

Also,

$$w(\neg\exists x\, \alpha_k(x)) = 1 - w(\exists x\, \alpha_k(x))$$
$$= 1 - \lim_{n\to\infty} w(\bigvee_{i=1}^{n}\alpha_k(a_i))$$
$$= 1 - \lim_{n\to\infty} w(\bigvee T_n^k)$$
$$= \lim_{n\to\infty}[1 - w(\bigvee T_n^k)]$$
$$= \lim_{n\to\infty} w(\bigvee U_n^k)$$

where U_n^k consists of all conjunctions

$$\alpha_{r_1}(a_1) \wedge \alpha_{r_2}(a_2) \wedge \ldots \wedge \alpha_{r_n}(a_n)$$

where none of the $r_j = k$. Similarly to the above , we have

$$\lim_{n\to\infty} w\left(\neg\exists x\,\alpha_k(x) \leftrightarrow \bigvee U_n^k\right) = 1.$$

It is now easy to see (or simply apply Lemma 3.7 from Paris & Vencovská, 2015) that

$$w\left(\bigwedge_{i=1}^{m} \alpha_{f_i}(b_i) \wedge \exists S \neg \exists \overline{S}\right) = \lim_{n\to\infty} w\left(\bigvee Y_n^S\right)$$

where Y_n^S consists of all conjunctions $\alpha_{r_1}(a_1)\wedge\alpha_{r_2}(a_2)\wedge\ldots\wedge\alpha_{r_n}(a_n)$ that are in T_n^k for all $k \in S$, in U_n^k for all $k \notin S$, and that imply $\bigwedge_{i=1}^m \alpha_{f_i}(b_i)$ (so n is large enough for the $a_1, a_2, \ldots, a_n$ to include all the b_i and if b_i is a_{s_i} then $r_{s_i} = f_i$). In other words all conjunctions which contain some $\alpha_k(a_j)$ for all $k \in S$, no $\alpha_k(a_j)$ for $k \notin S$, and all of the $\alpha_{f_i}(b_i)$ for $i = 1, 2, \ldots, m$.

By the same argument,

$$w\left(\bigwedge_{i=1}^{m} \alpha_{\sigma(f_i)}(b_i) \wedge \exists V \neg \exists \overline{V}\right) = \lim_{n\to\infty} w\left(\bigvee Z_n^V\right)$$

where Z_n^V consists of all conjunctions $\alpha_{r_1}(a_1) \wedge \alpha_{r_2}(a_2) \wedge \ldots \wedge \alpha_{r_n}(a_n)$ which contain some $\alpha_k(a_j)$ for all $k \in V$, no $\alpha_k(a_j)$ for $k \notin V$, and all of the $\alpha_{\sigma(f_i)}(b_i)$ for $i = 1, 2, \ldots, m$.

Now extend σ to a permutation of $\{1, 2, \ldots, 2^q\}$, so on $\overline{S}$, σ is a bijection to $\overline{V}$. Clearly the conjunctions in Y_n^S exactly pair off with those in Z_n^V via

$$\alpha_{r_1}(a_1)\wedge\alpha_{r_2}(a_2)\wedge\ldots\wedge\alpha_{r_n}(a_n) \leftrightarrow \alpha_{\sigma(r_1)}(a_1)\wedge\alpha_{\sigma(r_2)}(a_2)\wedge\ldots\wedge\alpha_{\sigma(r_n)}(a_n)$$

and by Ax each such pair have the same value under w. Finally since the conjunctions in Y_n^S are disjoint, and similarly for Z_n^V,

$$w\left(\bigvee Y_n^S\right) = \sum_{\Gamma \in Y_n^S} w(\Gamma) = \sum_{\Gamma \in Z_n^V} w(\Gamma) = w\left(\bigvee Z_n^V\right),$$

yielding (2) as required.

Extending this to general sentences using Lemma 2 gives that:

Theorem 1 *A probability function w on a unary language satisfies ENV iff w satisfies Ax.*

6   The Polyadic case

In view of Theorem 1 in the unary case we might have been inclined to conjecture that a probability function w on a polyadic language satisfies ENV iff it satisfies PIP.

ENV implies PIP so it would suffice if we could show that every probability function that satisfies PIP also satisfies ENV. Unfortunately this is not the case as we shall shortly demonstrate via a counter-example. Firstly however we show a partial result in this direction which informs us that PIP probability functions are invariant under automorphisms from a family of *special* exchangeable automorphisms which we now describe.

Let $\phi_1, \ldots, \phi_k$ be exclusive and exhaustive sentences of L without any constants and let τ be a permutation of $\{1, \ldots, k\}$. Let $F_1, \ldots, F_k$ be automorphisms of BL permuting atoms (so exchangeable) such that for each j,

$$\{F_j(M) : M \models \phi_j\} = \{M : M \models \phi_{\tau(j)}\},$$

equivalently,

$$M \models \phi_j \iff F_j(M) \models \phi_{\tau(j)}. \tag{3}$$

Then F defined on $\mathcal{T}$ by

$$F(M) = F_j(M) \text{ whenever } M \models \phi_j, \ \ j = 1, 2, \ldots, k,$$

that is,

$$F(\psi(\vec{a})) = F\left(\bigvee_j (\psi(\vec{a}) \wedge \phi_j)\right) = \bigvee_j F_j(\psi(\vec{a}) \wedge \phi_j) = \bigvee_j (F_j(\psi(\vec{a})) \wedge \phi_{\tau(j)}),$$

is an exchangeable automorphism.

An example of such a special automorphism for the language containing a single binary relation symbol R can be given as follows: let

$$\phi_1 = \exists x \forall y \, \neg R(x, y) \qquad \phi_2 = \forall x \exists y \, R(x, y)$$

and let τ be the identity. Let F_1 be the automorphism determined by the permutation of atoms which just transposes

$$R(x,x) \wedge R(y,y) \wedge R(x,y) \wedge \neg R(y,x), \ \ R(x,x) \wedge R(y,y) \wedge \neg R(x,y) \wedge R(y,x)$$

Jeff Paris and Alena Vencovská

and leaves the rest fixed (see Paris & Vencovská, 2015, page 297), and let F_2 be the identity. The only thing to check is that (3) is satisfied for F_1 but that is clear since if for some a_i, $M \models \forall y \neg R(a_i, y)$ then for this same a_i we must also have $F_1(M) \models \forall y \neg R(a_i, y)$.

A second example, given in (Paris & Vencovská, 2015) on page 301, again for this language, is with

$$\phi_1 = \forall x, y\, R(x,y), \quad \phi_2 = \forall x, y\, \neg R(x,y), \quad \phi_3 = \neg\phi_1 \wedge \neg\phi_2,$$

τ transposing 1 and 2, F_3 the identity and $F_1 = F_2$ the automorphism such that
$$F_1(M) \models R(a_i, a_j) \iff M \models \neg R(a_i, a_j).$$

For a special exchangeable automorphism F as defined above and w a probability function satisfying PIP we have

$$w(F(\psi)) = w(F\left(\bigvee_j(\psi \wedge \phi_j)\right)) = w\left(\bigvee_j(F_j(\psi) \wedge \phi_{\tau(j)})\right) =$$
$$\sum_j w(F_j(\psi) \wedge \phi_{\tau(j)}) = \sum_j w(F_j(\psi \wedge \phi_j)) = \sum_j w(\psi \wedge \phi_j) = w(\psi),$$

so PIP probability functions are invariant under automorphisms from the family of special exchangeable automorphisms.

Note that these special automorphisms actually satisfy an apparently stronger condition, namely that if δ is *any* map from $\mathbb{N}^+$ to $\mathbb{N}^+$ and

$$F(\theta(a_1, a_2, \ldots, a_n)) = \phi(a_{i_1}, a_{i_2}, \ldots, a_{i_m})$$

then

$$F(\theta(a_{\delta(1)}, a_{\delta(2)}, \ldots, a_{\delta(n)})) = \phi(a_{\delta(i_1)}, a_{\delta(i_2)}, \ldots, a_{\delta(i_m)}).$$

This follows since it holds for automorphisms which permute state descriptions.

A next step here might have been to show the result for exchangeable automorphisms satisfying the above condition. However the following example shows that there are automorphisms satisfying this condition for which there are PIP probability functions not invariant under them. So ENV, even if possibly weakened by imposing the above condition, is strictly stronger than PIP in the non-unary context.

Let L contain just two binary predicates R, Q. For $M \in \mathcal{T}$ define $F(M) \in \mathcal{T}$ to be the same as $F(M)$ except that if a_i, a_j are such that

$$M \vDash R(a_i, a_i) \wedge R(a_j, a_j) \wedge (\forall x\, Q(a_i, x) \vee \forall x\, Q(a_j, x))$$

then

$$M \vDash R(a_i, a_j) \wedge \neg R(a_j, a_i)$$

iff

$$F(M) \vDash \neg R(a_i, a_j) \wedge R(a_j, a_i).$$

Clearly F defines an exchangeable automorphism. Notice that

$$M \vDash \forall x\, Q(a_i, x) \iff F(M) \vDash \forall x\, Q(a_i, x).$$

Let $\Theta(a_1, a_2, a_3)$ be the state description represented by the pair of matrices

$$\begin{pmatrix} 1 & 1 & 0 \\ 0 & 1 & 1 \\ 0 & 0 & 1 \end{pmatrix}, \quad \begin{pmatrix} 1 & 1 & 1 \\ 1 & 1 & 1 \\ 1 & 1 & 1 \end{pmatrix}.$$

That is, the entry for ith row, jth column in the first matrix is

$$\begin{array}{ll} 1 & \text{if } \Theta(a_1, a_2, a_3) \vDash R(a_i, a_j), \\ 0 & \text{if } \Theta(a_1, a_2, a_3) \vDash \neg R(a_i, a_j), \end{array}$$

and similarly in the second matrix for Q. Let $\Theta'(a_1, a_2, a_3)$ be the state description represented by the pair

$$\begin{pmatrix} 1 & 0 & 0 \\ 1 & 1 & 1 \\ 0 & 0 & 1 \end{pmatrix}, \quad \begin{pmatrix} 1 & 1 & 1 \\ 1 & 1 & 1 \\ 1 & 1 & 1 \end{pmatrix}$$

and let

$$\chi_1 = \Theta(a_1, a_2, a_3) \wedge \forall x\, Q(a_1, x) \wedge \neg\forall x\, Q(a_2, x) \wedge \neg\forall x\, Q(a_3, x),$$

$$\chi_2 = \Theta'(a_1, a_2, a_3) \wedge \forall x\, Q(a_1, x) \wedge \neg\forall x\, Q(a_2, x) \wedge \neg\forall x\, Q(a_3, x).$$

Since $F(\chi_1) = \chi_2$, ENV mandates that these two sentences get the same probability. However, there are probability functions u satisfying PIP for which $u(\chi_1) > u(\chi_2)$ so PIP is strictly weaker than ENV.

To demonstrate such a probability function we shall assume familiarity with the notation, definitions and results given in (Paris & Vencovská, 2015) chapter 42. Let u be the probability function $u_{\bar{E}}^{\bar{p},L}$ satisfying PIP with

$$\bar{p} = \langle 0, 1/4, 1/4, 1/4, 1/4, 0, 0, \ldots \rangle,$$

$L = \{R, Q\}$ and $\bar{E}$ generated by $\langle 1, 2 \rangle \equiv_2 \langle 2, 3 \rangle$ (so $\langle 1 \rangle \equiv_1 \langle 2 \rangle \equiv_1 \langle 3 \rangle$ and $\langle 2, 1 \rangle \equiv_2 \langle 3, 2 \rangle$).

Note that $u(\chi_1), u(\chi_2)$ are the limits as n tends to infinity of sums of the $u(\Psi)$ over state descriptions $\Psi(a_1, a_2, \ldots, a_n)$ extending Θ and Θ' respectively and such that $\Psi \models Q(a_1, a_k)$ for every $k \in \{1, 2, \ldots, n\}$, $\Psi \models \neg Q(a_2, a_k)$ for some $k \in \{1, 2, \ldots, n\}$, and $\Psi \models \neg Q(a_3, a_k)$ for some $k \in \{1, 2, \ldots, n\}$. Let S_n and S_n' respectively denote the set of such extensions. Hence

$$u(\chi_1) = \sum_{\Psi \in S_n} u(\Psi) \quad = \sum_{\substack{\Psi \in S_n \\ \bar{c} \in \{1,2,3,4\}^n : \Psi \in \mathcal{C}_{\bar{E}}^L(\bar{c}, \bar{a})}} |\mathcal{C}_{\bar{E}}^L(\bar{c}, \bar{a})|^{-1} 4^{-n}$$

$$= \sum_{\bar{c} \in \{1,2,3,4,\}^n} \frac{|S_n \cap \mathcal{C}_{\bar{E}}^L(\bar{c}, \bar{a})|}{4^n |\mathcal{C}_{\bar{E}}^L(\bar{c}, \bar{a})|} \tag{4}$$

and similarly for $u(\chi_2)$.

Both $\Theta(a_1, a_2, a_3)$ and $\Theta'(a_1, a_2, a_3)$ have no repeated columns in their first matrices so the only vectors $\langle c_1, c_2, \ldots, c_n \rangle$ for which a Ψ from S_n or S_n' can appear in $\mathcal{C}_{\bar{E}}^L(\bar{c}, \bar{a})$ are those without repeats amongst the c_1, c_2, c_3. Moreover, all of the 1,2,3,4 have to appear amongst the $c_1, \ldots, c_n$ because otherwise there could be no k satisfying $\Psi \models \neg Q(a_2, a_k)$ and no k' satisfying $\Psi \models \neg Q(a_3, a_{k'})$. Clearly also for the first such k, k' we must have $k = k' = k_0$, say.

Notice that when all the 'colours' 1,2,3,4, have been used in $\bar{c}$, if Ψ is in $\mathcal{C}_{\bar{E}}^L(\bar{c}, \bar{a})$ and $c_{n+1} \in \{1, 2, 3, 4\}$ then Ψ has a unique extension to a state description in $\mathcal{C}_{\bar{E}}^L(\langle \bar{c}, c_{n+1} \rangle, \langle \bar{a}, a_{n+1} \rangle)$, and this extension is in S_{n+1} just when it is in S_n. Thus

$$\frac{|S_{n+1} \cap \mathcal{C}_{\bar{E}}^L(\langle \bar{c}, c_{n+1} \rangle, \langle \bar{a}, a_{n+1} \rangle)|}{\mathcal{C}_{\bar{E}}^L(\langle \bar{c}, c_{n+1} \rangle, \langle \bar{a}, a_{n+1} \rangle)|} = \frac{|S_n \cap \mathcal{C}_{\bar{E}}^L(\bar{c}, \bar{a})|}{|\mathcal{C}_{\bar{E}}^L(\bar{c}, \bar{a})|}.$$

Since there are always 4 possibilities for c_{n+1} and, as we shall shortly see, $S_4 \cap \mathcal{C}_{\bar{E}}^L(\bar{c}, \bar{a}) \neq \emptyset$ for $\bar{c} = \langle 1, 2, 3, 4 \rangle, \langle 3, 2, 1, 4 \rangle$, the contribution to (4) from $\vec{c}$'s extending $\langle 1, 2, 3, 4 \rangle$ or $\langle 3, 2, 1, 4 \rangle$, is non-zero. We will need this observation at the conclusion of this example.

Let $\Psi(a_1, a_2, \ldots, a_n) \in \mathcal{C}_{\vec{E}}^L(\vec{c}, \vec{a})$. We will consider which choices of $\vec{c}$ allow $\Psi \in S_n$ or $\Psi \in S_n'$.

First consider a colouring $\vec{c}$ such that $\langle c_1, c_2, c_3 \rangle$ is $\langle 2, 3, 4 \rangle$ (so $c_{k_0} = 1$). Here a_1, a_2, a_3, a_{k_0} have colours $2, 3, 4, 1$ respectively. For $\Psi \vDash \Theta$ we have $\Psi \vDash R(a_1, a_2) \wedge \neg R(a_2, a_1)$ where the colours are $\langle 2, 3 \rangle, \langle 3, 2 \rangle$ respectively. Hence for Ψ to also be in $\mathcal{C}_{\vec{E}}^L(\vec{c}, \vec{a})$ requires, since $\langle 1, 2 \rangle \equiv_2 \langle 2, 3 \rangle$, that $\Psi \vDash R(a_{k_0}, a_1) \wedge \neg R(a_1, a_{k_0})$. Furthermore since $\langle 1 \rangle \equiv_1 \langle 2 \rangle$ and $\Psi \vDash R(a_1, a_1)$ we must also have $\Psi \vDash R(a_{k_0}, a_{k_0})$. So to form $\Psi \in S_n \cap \mathcal{C}_{\vec{E}}^L(\vec{c}, \vec{a})$ the only free choices for the $R(a_i, a_j)$ concern $R(a_{k_0}, a_2), R(a_{k_0}, a_3)$.

An analogous argument for $\Psi \vDash \Theta'$ shows that there are again only the same free choices for R. Finally the choices for the $Q(a_i, a_j)$ are clearly the same for both Θ and Θ'. It follows that

$$|S_n \cap \mathcal{C}_{\vec{E}}^L(\vec{c}, \vec{a})| = |S_n' \cap \mathcal{C}_{\vec{E}}^L(\vec{c}, \vec{a})| \tag{5}$$

so $u(\chi_1), u(\chi_2)$ both gain the same contributions with these colourings.

Similar arguments give the same conclusion (5) for any colouring with $c_{k_0} \neq 4$. Note that (5) can be zero, for example when $\langle c_1, c_2, c_3 \rangle$ is $\langle 3, 2, 4 \rangle$ (so again $c_{k_0} = 1$) since any $\Psi \in \mathcal{C}_{\vec{E}}^L(\vec{c}, \vec{a})$ must satisfy $\Psi \vDash Q(a_2, a_{k_0}) \leftrightarrow Q(a_1, a_2)$ by virtue of $\langle 2, 1 \rangle \equiv_2 \langle 3, 2 \rangle$ but that precludes it from also satisfying $\neg \forall x Q(a_2, x)$.

Now consider $\vec{c}$ such that $\langle c_1, c_2, c_3 \rangle$ is $\langle 3, 1, 2 \rangle$ (so $c_{k_0} = 4$). Here a_1, a_2, a_3, a_{k_0} have colours $3, 1, 2, 4$ respectively. For $\Psi \vDash \Theta$ we must have $\Psi \vDash R(a_2, a_3) \wedge \neg R(a_3, a_1)$. But $\Psi \in \mathcal{C}_{\vec{E}}^L(\vec{c}, \vec{a})$ requires, since $\langle 1, 2 \rangle \equiv_2 \langle 2, 3 \rangle$, that $\Psi \vDash R(a_2, a_3) \leftrightarrow R(a_3, a_1)$. So no Ψ from $\mathcal{C}_{\vec{E}}^L(\vec{c}, \vec{a})$ contributes to S_n. The same conclusion follows similarly for S_n'. Indeed analogous reasoning shows that (5) holds and is 0 when $\vec{c}$ is such that $\langle c_1, c_2, c_3 \rangle$ is $\langle 1, 3, 2 \rangle, \langle 2, 1, 3 \rangle$ or $\langle 2, 3, 1 \rangle$.

In the case of $\langle 1, 2, 3 \rangle$, for Ψ to be in $\mathcal{C}_{\vec{E}}^L(\vec{c}, \vec{a})$ the conditions include $\Psi \vDash R(a_1, a_2) \leftrightarrow R(a_2, a_3)$ and $\Psi \vDash R(a_2, a_1) \leftrightarrow R(a_3, a_2)$. This is inconsistent with $\Psi \vDash \Theta'$ since $\Theta' \vDash \neg R(a_1, a_2) \wedge R(a_2, a_3)$ so $u(\chi_2)$ obtains no contribution from this colouring. However the conditions can be fulfilled for some $\Psi \vDash \Theta$ (with free choices for $Q(a_{k_0}, a_i), R(a_{k_0}, a_i), R(a_i, a_{k_0}), i = 1, 2, 3, k_0$) so in light of the observation we made earlier, this colouring provides a non-zero contribution to $u(\chi_1)$. The situation is the same for $\langle 3, 2, 1 \rangle$.

Summing over all choices of colouring then gives that $u(\chi_1) > u(\chi_2)$, as required.

7 Conclusion

In this paper we have considered strong symmetry principles in pure inductive logic. We have argued that the formulation of the Invariance Principle INV proposed in (Paris & Vencovská, 2015) should be amended and the resulting principle ENV, along with the ubiquitous Ex, should be considered as a candidate for the 'ultimate', most comprehensive representation of symmetry in pure inductive logic. We have shown that in the unary context ENV is equivalent to the well established principle Ax but for languages containing non-unary relation symbols, ENV + Ex is strictly stronger than the previously considered PIP + Ex.

References

Carnap, R. (1962). The aim of inductive logic. In *Logic, methodology and philosophy of science* (pp. 303–318). Stanford University Press.

Carnap, R., & Stegmüller, R. (1959). *Induktive Logik und Warscheinlichkeit*. Wien: Springer Verlag.

Johnson, W. (1932). Probability: The deductive and inductive problems. *Mind*, *41*, 409–423.

Paris, J. (2016). *A useful lemma in unary predicate logic*. http://oldwww.ma.man.ac.uk/ jeff/papers/useful.pdf.

Paris, J., & Vencovská, A. (2011). Symmetry's end? *Erkenntnis*, *74*(1), 53–67.

Paris, J., & Vencovská, A. (2015). *Pure Inductive Logic*. Cambridge: Cambridge University Press.

Paris, J., & Vencovská, A. (2017). *A note on automorphisms of the Lindenbaum algebra of SL and BL*. http://oldwww.ma.man.ac.uk/ jeff/papers/jp170829Lind.pdf.

Paris, J., & Vencovská, A. (2019). Translation invariance and Miller's weather example. *Journal of Logic, Language and Information*, *28*(4), 489–514.

Ronel, T., & Vencovská, A. (2014). Invariance principles in polyadic inductive logic. *Logique et Analyse*, *57*, 541-561.

Symmetry Principles in Pure Inductive Logic

Jeff Paris
University of Manchester, Department of Mathematics
United Kingdom
E-mail: jeffparis49@gmail.com

Alena Vencovská
University of Manchester, Department of Mathematics
United Kingdom
E-mail: alenavencovska1@gmail.com

Logical Pluralism and Genuine Logic

JAROSLAV PEREGRIN AND VLADIMÍR SVOBODA[1]

Abstract: The paper offers a critical reflection of the logical pluralism of Beall & Restall; urging the classical Carnapian pluralism that Beall & Restall tend to sideline. Our criticism rests on two claims a) There is no language in which we can formulate argument schemata, of which we are to decide whether they are valid or not. b) The very notion of correct or genuine logic is misleading. There are, we think, no criteria of correctness of an (alleged) system of logic beyond vague criteria of usefulness.

Keywords: logical pluralism; Beall; Restall; collapse argument

1 Introduction

Logic is a well-established scientific discipline with deep historical roots. Since ancient times it has been seen (though surely not by everyone) as the discipline that is meant to help us "reason effectively about practical affairs, stand his or her ground amid confusion, differentiate the certain from the probable, and so forth".[2] If we conceive of logic in this way, we will probably not be tempted to insist that the answers that logic provides should be always unanimous and certain. During the last one and a half centuries, however, logic has become closely intertwined with mathematics – a discipline in which we tend to expect unique definite answers to properly formulated questions (though it may be difficult to find them). Hence, it appeared quite natural to expect something similar of logic – to every truly logical question there should be a single correct answer. Those logicians who insist that the quest for such definite answers makes sense or is even imperative for logical studies are commonly classified as *logical monists*, while those who think

---

[1] Work on this paper was supported by Grant No. 20-18675S of the Czech Science Foundation and coordinated by the Institute of Philosophy of the Czech Academy of Sciences in Prague. The authors are grateful to Vít Punčochář for valuable critical comments.

[2] This particular conception of the mission of logic is ascribed to the Greek Stoics (see Inwood, 2003, p. 229). So as not to distort the historical picture, we should note that the Stoics were perhaps more practically oriented than most philosophers and logicians of the two millennia to come after their era.

that we should not be so categorical and that it is preferable to open space for embracing more different logics are called *logical pluralists*.[3]

The debate between the monists and the pluralists wasn't very vigorous during the concluding decades of the 20th century, but it received new momentum at the beginning of the present century after Jc Beall and Greg Restall published their original defense of logical pluralism (see esp. Beall & Restall, 2000 and 2006). Their specific presentation of the issue and their original argumentation received considerable attention and the subsequent debate remains ongoing.

This paper presents a contribution to the debate. We, on the one hand, intend to uphold logical pluralism and, in this respect, we are in the same boat as Beall & Restall, but we are critical of their specific version of logical pluralism – we suggest that the framing of their project as an alternative to Carnapian pluralism is misconceived. Our critique is meant to reinforce the classical Carnapian pluralism that Beall & Restall tend to sideline. (They, for sure, don't explicitly deny classical pluralism, but what they say indicates that they view their version of pluralism as being more substantial than the plain Carnapian version.) Our main motive, however, is more general. We don't just want to defend a Carnapian stance, but additionally to promote a general picture within which logic is not (at least not principally) a kind of (meta)mathematical theory (or, for that matter, a complex of such theories) but rather a multifaceted tool that is meant to help us reason effectively about practical as well as theoretical matters and to prevent and resolve confusion.

2 Logical pluralism of Beall and Restall

As we have suggested, Beall and Restall (2000, 2006, hereafter B&R) breathed new life into the long-standing discussion about logical pluralism, in particular the debate over the question of whether there is only one correct logic, or whether there are more, equally correct, ones. It is important to stress that their variety of logical pluralism is meant to cut essentially deeper than pointing out the trivial fact that logicians study many different logical systems. B&R seem to take the obvious plurality of systems of logic as a fact which is not necessary to mention. They, however, think that there is something important concerning the plurality of logics that must be pointed out. In their eyes, it is important to contrast the specific version of pluralism

---

[3]The two categories don't cover all those dealing with such issues, we may consider also different positions, such as logical nihilism (Russell, 2018).

they advocate with the Carnapian one, which is a matter of the principle of tolerance: "In logic, there are no morals. Everyone is at liberty to build his own logic, i.e. his own form of language, as he wishes. All that is required of him is that, if he wishes to discuss it, he must state his methods clearly, and give syntactical rules instead of philosophical arguments" (Carnap, 1937, p. 52). Beall and Restall (2006, p. 78-9) write: "What we want to emphasise is that Carnap's pluralism is not our kind of logical pluralism. .... For us, pluralism can arise within a language as well as between languages." They claim that there are several versions of the relation of logical consequence, all of which (i) are equally correct and (ii) are hosted by the same language.

B&R adopt, as the basis for their deliberations, a conception of logical consequence that, though not universally adopted, can surely be classified as the mainstream among contemporary logicians. They formulate their argumentation in terms of what they call the *Generalized Tarski Thesis*:

> (GTT) An argument is valid$_x$ if and only if, in every case$_x$ in which the premises are true, so is the conclusion. (Beall & Restall, 2006, p. 29)

(The suggested perspective on logical consequence may not be as a matter-of-course as B&R seem to suppose, but it prevails among today's logicians and we don't want to challenge it here.) An argument, according to B&R, is valid iff it preserves truth in every case; *viz.* if there is no case when its premises are true and the conclusion false. The crucial question that needs to be addressed, according to them, is *What are the cases*? We already indicated what the outcome of their deliberation is – they claim that there is no one correct answer to this question, that it can be answered in more than one way depending on how we exactly specify the concept of case. Consequently, they argue, we have several logics all of which are correct and hence we can't and needn't choose among them – we can embrace any of them.

B&R are more specific about the outcome – GTT yields us at least three respectable logics: *Classical logic* (CL) which we receive if we take the cases to be Tarskian models or possible worlds, *Intuitionistic logic* (IL) which we receive if we take the cases to be constructions, and *Relevant logic* (RL) that results from taking the cases to be situations.

Let us consider, for the sake of illustration, two simple argument schemas. The schema of double negation:

(DN) $\dfrac{\neg\neg A}{A}$

and the schema of explosion:

(EX) $\dfrac{A \quad \neg A}{B}$

The first schema is, as we know, valid from the perspectives of classical logic and relevant logic, while being invalid from the perspectives of intuitionistic logic, whereas the second one is valid in classical and intuitionistic logics, while not being valid in relevant logic.

B&R claim that as there is no way to decide which specification of the concept of case is the proper one, there is no way to adjudicate among the different logics they yield. Hence there is no one correct logic, there are more which are equally correct – which meet the requirements that we have regarding the concept of logical consequence.

3   The collapse argument

Arguments against the viability of this kind of pluralism were raised by several distinguished scholars, most notably by Stephen Read (2006) and Rosanna Keefe (2014) (hereafter R&K).[4] They claim that if an argument is sanctioned as valid by one of the "correct" logics, then it must be seen as *valid simpliciter* (no matter what status argument has from the viewpoint of alternative versions of logical validity). The point is that once we know that an argument is *valid* in some respectable logic, we know that it is guaranteed to lead us from true premises to a true conclusion; and we need not be bothered that it is not valid in other logics. (We should keep in mind that being not valid in a logic is not necessarily being *invalid* in the sense that an argument is guaranteed to lead us from some true premises to a false conclusion.)[5]

Keefe (2014, p. 1385) formulates the argument that in her view all logicians accepting B&R's principles should take into account as follows:

---

[4] Read (2006) points out that the argument appears (in not so clearly articulated form) already in (Priest, 2001).

[5] Cf. (Svoboda & Peregrin, 2016).

Consider, next, a relatively ordinary context of reasoning, and suppose our subject, S, endorses Beall and Restall's pluralism. If S accepts premises Γ and is considering conclusion C, what logic should she call on to decide whether or not to accept C? I argue that she ought to endorse the argument and accept its conclusion if it is valid according to any of the acceptable relations in the plurality. Suppose "Γ therefore C" is valid on some acceptable logics and not on others: does the truth of Γ guarantee the truth of C? Yes, and no, because it depends what you mean by "guarantee", and there is no unique sense to the claim that the conclusion is guaranteed to be true. But, no sense of "guarantee" is compatible with the premises being actually true and the conclusion actually false, even if there is variation over the putative cases the truth-transmission must travel across. So, if the argument is valid in any sense, that is enough to show that the conclusion is actually true, assuming the premises are.

And as all three versions of logical validity considered by B&R are *logical* in the sense that they cannot actually lead us from true premises to a false conclusion, it follows that (for example) both (DN) and (EX) are *valid simpliciter*. B&R's pluralism thus inevitably collapses – there is always just one answer to the question whether an argument (form) is valid.

This argumentation evoked a discussion which perhaps hasn't come to an end yet (see, e.g., Caret, 2017; Kouri Kissel & Shapiro, 2020; Stei, 2020). We are not interested in its details, we just want to point out that insofar as it accepts the stage-setting of B&R, it is also liable to our criticism.

4 Which language?

The collapse argument may be seen as generally convincing,[6] but it is (or at least may be) somewhat misleading. The core of the possible problem is relatively simple. R&K tacitly accept B&R's assumption that when we consider the relation of logical consequence we may deal with *different evaluations* of (the validity of) the *same arguments*. This is by no means striking as B&R do insist that the three differing logics in question provide,

---

[6]A more cautious approach would be to speak about collapse *arguments*, as individual authors differ in details which can perhaps be relevant for a proper assessment of the individual versions.

in some cases, different assessments of the very same arguments. But we have serious doubts about it.

Let us again consider the argument – more precisely an argument form[7] – (DN):

(DN) $\dfrac{\neg\neg A}{A}$

The question we want to press is the following: *In which language are arguments which are instances of this form formulated?* We can see several possible answers to the simple question:

  (i) In a natural language (like English).

 (ii) In an uninterpreted artificial language.

(iii) In an interpreted artificial language.

(iv) In a 'semi-interpreted' artificial language.

 (v) In a language beyond natural and artificial languages.

Let us consider the individual answers one by one.

5 Natural language?

Could it be that schemas like (DN) or (EX) have as their instances formations of sentences of a natural language? In such a case the symbols "$\neg$" and "$\vee$" would have to be shortcuts for certain expressions of the natural language, let us say, English – most plausibly for "it is not the case that" and "or". Then, when we consider the (in)validity of (DN) we face questions like *Are arguments of the form*

$\dfrac{\textit{It is not the case that it is not the case that } A}{A}$

correct in English?

---

[7]It is important to distinguish between *arguments* and *argument forms* or *schemata*. While this is a very clear distinction, it is often obscured, which contributes to confusions that are easy to overlook.

What knowledge do we need so that we can properly answer this question (or the questions in which a meaningful English sentence is substituted for A)? The answer is, in a sense, easy – we need to know what exactly "it is not the case that" means in English, how exactly it contributes to the meaning of the sentences in which it occurs (and which arguments its presence thus supports). This, obviously, is a question for empirical linguists, not for logicians. Such a question may be answered by empirical research among competent speakers. The research might, we contend, provide some interesting insights but it is unlikely that we would receive a clear and definite answer to our question and it would be similar (we estimate) for other natural languages. It is clear in any case that this is not a question that logicians would be competent to answer.[8]

6   An uninterpreted artificial language?

If the language of the relevant instances is not a natural one, could it be an artificial one of the kind produced by logicians? Could it be, e.g., something like the language of the first-order predicate calculus? Before we try to answer this we must disambiguate the term "language": sometimes it means just a syntactic system (a set of primitive symbols plus formation rules), in other cases it includes a semantics (be it model-theoretical or proof-theoretical, or perhaps yet another one).

In (Peregrin & Svoboda, 2017) we distinguished between *formal* languages, which consist of interpreted logical constants plus uninterpreted parameters, and *formalized* languages, which consist of interpreted (logical and extralogical) constants. Here, since what is in question is the interpretation of logical constants, both of these kinds of languages would count as interpreted, what we take here as uninterpreted languages (or *bare* languages) are languages in which nothing, not even logical constants (or, more precisely, what might become them) have as of yet fixed meanings. In the present section we will consider the first alternative, hence the possibility

---

[8]Note that the same holds when (DN) is articulated in an artificial language, the purpose of which is to mirror natural language as closely as possible. The point is that in such a case the validity of (DN) will again derive from facts about the natural language, to be discovered by an empirical research. And if somebody insists that an expression of an (uninterpreted) artificial language deserves to be called negation because its intended function is to serve as a means of such a mirroring, then it is the same case again.

that the formulas featuring in the schemas (DN) and (EX) are formulas of a
bare artificial language, i.e. not sentences but rather bare sentence forms.[9]

Given that the intended cases, resp. "instances", are formulas of this sort
it is obvious that there is a space for assigning different meanings to the
symbols that are to function as logical constants of the language – a language
that is clearly not yet classical (nor intuitionistic, nor relevant ...). But this
does not work for obvious reasons. Given the symbols of the language including those that are to function as logical constants are uninterpreted, it
is misleading to use the standard signs indicating a certain reading of the
formulas. In this way (DN) is misleading. We, in fact, face the question of
the following kind: *Should we take*

(DN) $\dfrac{\#\#A}{A}$

as valid?

It is clear that such a question does not make much sense – there is no
reason to answer it in the positive or in the negative until we are given a
guidance concerning the way in which "#" contributes to the meaning of
the sentences in which it occurs. But if (DN) is really articulated in a bare
language we "by definition" can't expect any such guidance. So we conclude
that the variant (ii) is out of consideration too.

7 An interpreted artificial language?

Let us proceed and consider the possibility that (DN) consists of formulas of
an *interpreted* artificial language. (As we are interested in logical constants
only, what makes a language interpreted for us is the interpretation of the
constants. Both our formal and formalized languages are of this kind.) This
alternative, however, straightforwardly leads to an uninteresting outcome:
as the constants of such a language are interpreted, we have nothing like a
"language of propositional logic", but only of *classical* (or *intuitionistic* or
relevant or ...) propositional logic. But once we deal with such a specific
language, the correctness of the relevant arguments which are instances of the

---

[9]As the formulas of a bare language are not sentences, but rather uninterpreted formulas,
here the instances of forms are again forms. But let's neglect this problem, as we often swallow
conventions when we theorize that are even less plausible. After all, we don't have to insist on
an intuitively plausible notion of instance – we can admit that we use the term technically.

schema can hardly be seen as an open issue. In fact, here we face questions of the kind of *Should we take*

(DN) $\dfrac{\neg\neg A}{A}$

(*where the negation is classical*) *as valid*?

And it is clear that this question is answered once we specify the interpreted language in question – if it is that of classical logic, for example, then the schema is valid; if it that of intuitionistic logic, it is not. An interpreted artificial language is thus out of consideration too.

8   A semi-interpreted artificial language?

Given that neither an uninterpreted, nor an interpreted, language is what can play the role of the language suitable for articulation of (the instances of) (DN), we might perhaps try a kind of a middle way. Maybe we can have a language with some partial interpretation, a language interpreted enough to make the symbol "¬" into a negation, but not yet the *classical* (or *intuitionistic* or *relevant* or ...) negation (and similarly for other logical operators). In fact, what B&R say indicates that this is what they are after: "The case is starker, of course, when it comes to classical, relevant, and intuitionistic logic, where arguments in *the one formal language (the language of conjunction, disjunction and negation,* for example) yield different verdicts of validity." (Beall & Restall, 2006, p. 79; our emphasis)

Do B&R presuppose that along with the well-known formal languages of classical, intuitionist and relevant logics, we should also consider "the formal language of conjunction, disjunction and negation"? Is this supposed to be a language that is somehow partially interpreted so that one of its constants is a negation, but not yet a negation of one of the specific logical systems? Should we see it as a language of "generic" propositional logic?

Be it as it may, we do not see any language of this kind and B&R don't provide any useful guidance. Thus we cannot but speculate. One option seems to be that what B&R have in mind when they speak about the common formal language is a logical language that employs some "minimal" connectives stripped off all the characteristics that determine whether the connective in question is one that fits into classical, intuitionist or other logical theory. It is, however, hard to believe that they would be after such connectives and yet they wouldn't try to be more specific.[10]

It may seem that we (at least we logicians) do use terms like "negation" to refer to expressions (or concepts) that need not be any specific kinds of negations, yet are already negations. Hence, it may seem that there is something as a negation *per se*, a generic negation that is not yet classical or intuitionistic, etc., there *must* be something like this, for it is what the term – in the professional jargon of logicians – refers to. But in our view, this is futile.

In fact, we use such terms as "negation" in a rather promiscuous way: sometimes to refer to a specific, e.g., classical, negation, sometimes to some expression that we intend to treat as a negation (a means of denying), sometimes perhaps to a word or a construction that, in a natural language, functions as a paraphrase or a translation of English "not" (which, for the speakers of English, is the ultimate prototypical means of negation) and perhaps in other ways. Nevertheless, as far as we can see, none of the common uses of the term "negation" individually substantiates the conviction that the term also denotes a "generic" negation belonging to a generic logical language. Someone might suggest that all the uses taken together delineate the required generic notion of negation but it is, we suggest, an illusion which is due to our unconscious tendency to suppose that if we (seem to) understand each other when we use a certain term, there must be something that the term denotes.

9 A language beyond or behind natural and artificial languages?

Now we have reached the last of the alternatives we can think of – the alternative that schemas like (DN) or (EX) (or their instances), i.e. schemas that are suited to come out as valid from the perspective of one logical theory and as invalid from that of another one, belong neither to a natural nor to an artificial language. So is there any other language to which they can belong?

There is of course a long tradition of considering languages that are not produced by us, fallible humans, but which are, as it were, bestowed

---

[10]To avoid misunderstanding: we do not claim that such a semi-interpreted language is impossible, and even not that it does not exist (cf. Punčochář, 2019, §3). We just claim that to assess whether it provides a satisfactory account of the connectives, it would have to be discussed in detail.

on us by a god, by nature itself, etc. Hence, the idea is that there is a language independent of us, one which we may only try to approximate by our imperfect languages. What kind of language might this be – where is it to be sought?

There are two prominent possibilities where such a language can reside. According to the first one, the language should be sought in human mind – not in an individual one, but as a kind of universal structure determining proper thought. Probably the most popular way of elaborating this idea leads us to consider the "language of thought" (LOT) – a specific kind of a structure that governs individual thought but transcends individual minds in the way the Husserlian *transcendental ego* does.

The idea of such a language is in many respects appealing – it seems quite plausible that when we talk to each other we convey thoughts of which we can think as of sentences of a (or the) LOT. But despite decades of philosophers musing about it,[11] it is not clear how to get hold of it, not only to learn about its properties, but even to see that it really exists. So we don't find any plausible substantiation for adoption of the idea that the relevant (instances of the) schemas like (DN) and (EX) should be seen as composed of items belonging to a LOT.

The other possibility is to situate the language in which (DN) and (EX) are articulated in a kind of Platonist heaven. This picture, unsurprisingly, does not satisfy us either. If we want to see mathematical objects and structures as residing in such a heaven, that is fair enough. However, such a heaven harbors all kinds of structures that can be seen as languages (or at least "languages") along with an immense number of structures that cannot be seen even as "languages" in scare quotes. How can we identify the language which hosts (DN) – the one to be used to express arguments and to be studied by logic?

In general, we are not convinced that we are justified in assuming the existence of a language beyond our natural languages and the artificial languages we have put together. Moreover, we don't believe that to elucidate the nature of our logic we *need* any such assumption.[12]

---

[11] See especially (Fodor, 1975, 2008).

[12] See (Peregrin & Svoboda, 2021) for a thorough discussion.

10   No language, no logic

Let us briefly recapitulate how we reached what looks like a dead end regarding our considerations about what B&R have in mind when talking about negation (or conjunction, disjunction, implication, ...) belonging to a language common for CPL, IL, RL (and perhaps also for other logics). We have considered five different answers to the question that asks in what kind of "generic" language the schemas (DN) and (EX) might be articulated. We eliminated the first three of them as evidently implausible. The other two are different – they both seem to open some room for speculation, but B&R do not seem to give us any clear direction towards a plausible answer to our simple question, and this is frustrating.

What is left? Perhaps we overlooked some answer to our question. The other possibility is that our question cannot be (for some reason which escapes us) answered at all. We are, in fact, afraid that the last possibility is quite likely, as not one of the witty commentators on B&R's version of pluralism – to our knowledge – challenged their assumption that there is a common formal language which the different logics share. But we will not be satisfied until it is clear what kind of language it is[13].

We think it is good to keep things as simple and perspicuous as possible. This leads us to suggest that there are no good reasons to suppose that there are languages beyond our natural languages and the artificial languages we have created by our definitions and conventions. We thus propose to forget about them. And as there is no logic without a language, we assume that there is no logic beyond those embodied in our natural languages and in our artificial languages. As we have seen, we have good reasons to conclude that no such language is suited to harbor a single "genuine" logic and no one is suited to harbor a "generic" logic. We thus dare conclude that there are no such logics.

We do not want to claim that logicians don't face dilemmas which involve choice among versions of logical connectives. There are different kinds of situations in which a logician can vacillate whether certain types of arguments that appear to have the structure of, to stick to our examples, (DN) or (EX) should be classified as valid, i.e. whether, say, classical logic with its negation is the right choice or whether intuitionist logic (with its negation) is preferable

---

[13]Recall the principle of tolerance: Everyone is at liberty to build his own form of language, all that is required is that "if he wishes to discuss it, he must state his methods clearly, and give syntactical rules instead of philosophical arguments". What we miss are syntactical rules specifying the (alleged) language in question.

in the given context. This is, we are convinced, the process of calibrating artificial languages *vis-à-vis* the natural one in order for the former to be usable for the purposes of regimentation and analysis of the latter.

B&R are open to accepting several mutually incompatible logical theories as true logics – logics that are guaranteed to be correct and among which we are free to choose. Others, who share their intuition that it is reasonable (or even necessary) to hypothesize "the formal language of conjunction, disjunction and negation" in which schemas like those on which we have focused are formulated, will have an urge to find out whether they are really valid or not. We, similarly as B&R, deny that there is a way to find out whether (DN) "really" holds or not.[14] However, what *we* deny is that there is (DN) as such (and that there is a formal language to which this (DN) as such belongs) – the only legitimately formed versions of (DN) are those belonging to a certain language, to a natural language or to a language of logic, be it that of classical logic, or intuitionistic logic, or some other kind of logic.

The point of logical studies is, we can say very crudely, *constructing* languages that allow us to formulate different meaningful versions of the pre-theoretical (DN) which English speakers identify as that which is common to the arguments like

(DNEng1)   *It is not the case that John is not smart*
             John is smart

(DNEng2)   *It is not true that Trump is not egoistic*
             Trump is egoistic

(DNEng3)   *It is not the case that mammals don't fly*
             Mammals fly

(DNEng4)   *It is false that 224 is not divisible by 7*
             224 is divisible by 7

---

[14] Here we have focused on (DN) to keep things simple. A wide variety of illustrative examples arises in connection with the connective (or rather bunch of connectives) identified as *implication*. It is this connective which was at the center of the ancient Stoic debates that anticipated the modern disputes concerning (the potential) plurality of logics, as it turned out that the meaning of the phrase that characterizes the most common conditionals can be explicated in different ways, e.g., in the way favored by Chrisippus or the way favored by Philo (cf. Kneale & Kneale, 1962, Chapter 3).

Apparently, all these arguments have something in common. What they have in common is, we can say, their logical form. We try to materialize this abstract *e pluribus unum* as a formula. We may, from the beginning, want to restrict ourselves to the means of a specific logical system (like classical propositional logic), which suggests to us a concrete version of negation (while we can use more complex analyses to account for more complex kinds of negation – like Russell's celebrated analysis of "The king of France is not bald".)

Another possibility is that we do not restrict ourselves to one logic and, along with seeking the most suitable formula to materialize the logical form, we also seek the logic from which the formula can come. In this case we have, already at the beginning, more versions of negations to choose from.

11   Conclusion

What B&R claim is that there is no single, correct logic; there are, in fact, at least three. R&K oppose this by saying that if there are three, there is only one. We argue that this way of framing the pluralism/monism debate may be misleading. What we object to in B&R (and also in R&K, insofar as they adopt B&R's framework) is that:

a) There is no language in which we can formulate schemata like (DN) or (EX), of which we are to decide whether they are valid or not. (What we can do is to set out to explicate, e.g., English negation and decide whether it is classical, intuitionistic or whatever negation that is up to the task – but the pluralism presupposed by this enterprise is of the Carnapian variety.)

b) The very notion of correct or genuine logic is misleading (independently of whether one insists that there is only one or there are more). There are, we think, no criteria of correctness of an (alleged) system of logic beyond vague criteria of usefulness.[15]

We are convinced that attempts at pinpointing the correct logic (or for that matter more corrects logics) are futile. There is nothing like "genuine" logical constants and nothing like a "genuine" logical consequence. (In fact, B&R themselves come close to this standpoint when stating: "Logic, whatever it is, must be a tool useful for the analysis of the inferential relationships between premises and conclusions expressed in arguments we actually employ", Beall & Restall, 2006, p. 8.)

---

[15] See (Peregrin & Svoboda, 2022) for a thorough discussion.

We see the project of B&R as opening an interesting vista on the problem of logical pluralism but in the end as somewhat misconceived. And we, for the same reasons, suggest that the debate on the collapse argument is misconceived too (though it also raises interesting issues). Both B&R's project and the debate are based on the idea that we can put aside the Carnapian pluralism and identify a true pluralism which is more authentic. The idea is, in our view, a seductive but potentially misleading illusion. In our view, we can form a Carnapian multitude of artificial languages that are meant to help us overcome (to some extent) the ambiguity and indistinctness of natural languages. There is, however, no artificial language that we could pick up and say: *"This is the language where the serious business of logic should be done. Let us find out whether it harbors only one logic, or more."*

What is closest to a "genuine" logic for a person is the logic implicit to the natural language that she uses to argue and reason. But this is a mere "protologic", which

i) is not articulate enough to act itself as logic in the sense adopted in contemporary philosophy; and

ii) may slightly vary among different natural languages (i.e. persons with different mother tongues can have slightly different negations, etc.).

Therefore, we must create our artificial languages as a means of its commonly acceptable regimentation in a process of zooming in on a reflective equilibrium: of turning the "protologic" into a (certain) logic proper. There are various ways of doing this, so here there is enough space for the Carnapian pluralism but not for a kind of "more genuine" pluralism.

References

Beall, J. C., & Restall, G. (2000). Logical pluralism. *Australasian journal of philosophy*, *78*(4), 475–493.

Beall, J. C., & Restall, G. (2006). *Logical Pluralism*. New York: Oxford University Press.

Caret, C. R. (2017). The collapse of logical pluralism has been greatly exaggerated. *Erkenntnis*, *82*(4), 739–760.

Carnap, R. (1937). *The Logical Syntax of Language*. London: Routledge.

Fodor, J. A. (1975). *The Language of Thought*. Cambridge (Mass.): Harvard University Press.

Fodor, J. A. (2008). *LOT 2: The Language of Thought Revisited*. Oxford: Oxford University Press.

Inwood, B. (2003). Stoicism. In *Routledge History of Philosophy Volume II* (pp. 243–273). London: Routledge.

Keefe, R. (2014). What logical pluralism cannot be. *Synthese*, *191*(7), 1375–1390.

Kneale, W., & Kneale, M. (1962). *The Development of Logic*. Oxford: Oxford University Press.

Kouri Kissel, T., & Shapiro, S. (2020). Logical pluralism and normativity. *Inquiry*, *63*(3-4), 389–410.

Peregrin, J., & Svoboda, V. (2017). *Reflective Equilibrium and the Principles of Logical Analysis: Understanding the Laws of Logic*. New York: Routledge.

Peregrin, J., & Svoboda, V. (2021). Moderate anti-exceptionalism and earthborn logic. *Synthese*, *199*(3-4), 8781–8806.

Peregrin, J., & Svoboda, V. (2022). Logica dominans vs. logica serviens. *Logic and Logical Philosophy*, *31*(2), 183-207.

Priest, G. (2001). Logic: One or many. In J. Woods & B. Brown (Eds.), *Logical Consequences: Rival Approaches* (pp. 23–38). Oxford: Hermes Scientific Publishers.

Punčochář, V. (2019). Substructural inquisitive logics. *The Review of Symbolic Logic*, *12*(2), 296–330.

Read, S. (2006). Monism: The one true logic. In D. Devidi & T. Keynon (Eds.), *A Logical Approach to Philosophy: Essays in Honour of Graham Solomon* (pp. 193–209). Cham: Springer.

Russell, G. (2018). Logical nihilism: Could there be no logic? *Philosophical Issues*, *28*(1).

Stei, E. (2020). Rivalry, normativity, and the collapse of logical pluralism. *Inquiry*, *63*(3-4), 411–432.

Svoboda, V., & Peregrin, J. (2016). Logically incorrect arguments. *Argumentation*, *30*(3), 263–287.

Jaroslav Peregrin
Czech Academy of Sciences, Institute of Philosophy
The Czech Republic
E-mail: peregrin@flu.cas.cz

Vladimír Svoboda
Czech Academy of Sciences, Institute of Philosophy
The Czech Republic
E-mail: svobodav@flu.cas.cz

Four Constructivist Attitudes in Prawitzian Semantics

ANTONIO PICCOLOMINI D'ARAGONA[1]

Abstract: I argue that some well-known alternative ways for developing Prawitz-inspired semantics are not coincidental, but respond to two basic dualities. This stems from the fact that what one has to evaluate in Prawitz's semantics is, not only the meaning of the components of one's alphabet, but also the acceptability of generalised eliminations over meaning-determination. In turn, this depends on the fact that Prawitz's semantics can be understood as a generalisation of Prawitz's own normalisation theory. We thus have at least four pairwise "symmetric" Prawitzian semantics, each amounting to a potential attitude towards Prawitz's constructivism. Thus, Prawitz's semantics can be understood as a conceptual and formal grid where to articulate harmoniously proof-based semantics accounting for the interplay between meaning determination and justification of deduction.

Keywords: Constructivism, proof, monotonicity, schematicity

1 Introduction

Prawitz's semantics (Prawitz, 1971, 1973) is an example of proof-theoretic semantics (Francez, 2015; Schroeder-Heister, 2018) where proofs are understood as valid arguments. An argument is in turn understood as a sequence of formulas arranged in a Gentzen's natural deduction tree-form style. Each node of the tree corresponds to an arbitrary inference, coming with an (alleged) justification of its validity.[2]

It is known that Prawitz's semantics can be articulated in many alternative ways, depending on the choices one makes relative to fundamental notions

---

[1]I thank Bruno Bentzen, Cesare Cozzo, Ansten Klev, Thomas Piecha, Peter Schroeder-Heister, Göran Sundholm and Will Stafford for illuminating discussions. Work on this article was supported by grant LQ300092101 from the Czech Academy of Sciences

[2]In a more recent development of Prawitz's semantics, the notion of valid argument is replaced by that of (epistemic) ground (Prawitz, 2015). Although the two approaches obviously intersect, they nonetheless differ on many points – see (Piccolomini d'Aragona, 2022a). What I will be saying here can in any case be applied to the theory of grounds too.

such as local validity – i.e. validity of an argument over a specific domain – or global validity – i.e. logical validity of the argument. What I want to highlight here is that these alternatives are not coincidental, but respond to some "symmetries". In particular, the alternatives can be regrouped in such a way as to give rise to a double duality, a *conceptual* one and a *structural* one. One may then wonder whether the poles of the dualities provide a complete and ordered classification of potential semantics in line with Prawitz's constructivism, a topic I have dealt with elsewhere (Piccolomini d'Aragona, 2022b). Below, I shall adopt a different viewpoint, and argue that the poles amount to four possible, pairwise intertwined "attitudes" for conceiving of Prawitz's tenets: *semantic vs deductive* attitude (in the conceptual duality) and *continuist vs discontinuist* attitude (in the structural duality).

I will also maintain that the reason of such a complex overall articulation of Prawitz's semantics is, roughly speaking, that the latter can be seen as a generalisation of Prawitz's normalisation theory for Gentzen's natural deduction. More precisely, Prawitz's normalisation theorems imply a number of side-results which are liable to a semantic reading, as they seemingly *confirm* Gentzen's idea that introductions fix the meaning of the logical constants, and thereby justify an *endorsement* of Dummett's *fundamental assumption* that provability is tantamount to provability by introduction. The linchpin between the confirmation and the endorsement is given by Prawitz's *inversion principle*, according to which proofs ending by introduction already contain the information needed for drawing consequences from their conclusions. This provides a criterion for a semantic justification of generalised eliminations, which ends up interacting with the interpretation of the non-logical terminology involved in proofs. The aforementioned dualities stem precisely from the interplay between non-logical meaning and justification of inferences.

I will conclude by raising some questions about whether one can make sense, not only of the interaction between the poles of the dualities, but also of the interaction between the dualities themselves, in particular by introducing a further semantic ingredient besides non-logical vocabulary and generalised eliminations, i.e. order of inferences in proofs.

2   Two semantic insights of normalisation

Prawitz's normalisation theorems for systems in a Gentzen's natural deduction format (Gentzen, 1935; Prawitz, 1965) show that derivations for *A* from

Γ reduce to *normal* derivations for A from $\Gamma^* \subseteq \Gamma$. A derivation is said to be normal when it contains no *detours*, namely, occurrences of formulas which are both conclusions of introductions and major premises of eliminations. Reducibility obtains by repeatedly appliying certain *reduction functions*, e.g.

$$
\cfrac{\cfrac{\begin{matrix} 1 \\ [A] \\ \mathscr{D}_1 \\ B \end{matrix}}{A \to B}\,(\to_I),\,1 \qquad \begin{matrix} \mathscr{D}_2 \\ A \end{matrix}}{B}\,(\to_E) \qquad \text{reduces to} \qquad \begin{matrix} \mathscr{D}_2 \\ [A] \\ \mathscr{D}_1 \\ B \end{matrix}
$$

Prawitz's theorems, and many of the corollaries they imply, are seemingly liable to a promising semantic reading. In what follows, I shall restrict myself to propositional logic, and be mainly concerned with two aspects.[3]

First, normalisation remains provable when one's logic is meant to act upon an underlying *atomic system*. Atomic systems can be defined in various ways, but here we can content ourselves with understanding them as (recursive) sets of *production rules*

$$
\frac{A_1, ..., A_n}{B}
$$

where A_i, B are atomic formulas of one's background language, and $A_i \neq \perp$ $(i \leq n)$.[4] Thus, normalisation can be understood as a structural property of logical derivations which is invariant over atomic derivability. And this is

---

[3]Other semantically significant consequences of Prawitz's normalisation, which I cannot deal with here, are for example the *normal-form theorem*, according to which paths in normal derivations consists of applications of eliminations, followed by applications of atomic rules or rules for $\perp$, followed by introductions, or the *sub-formula principle*, according to which the formulas involved in a normal derivation for A from Γ are sub-formulas of A or of one of the elements in Γ. Moreover, it is not difficult to see that Prawitz's detours, reduction functions and normalisation are nothing but the natural deduction version of, respectively, β-redexes, β-reductions and β-normalisation in typed λ-calculus. This is obvious from the Curry-Howard isomorphism, as is obvious that Prawitz's normalisation can be conceived of as the natural deduction version of Gentzen's *Hauptsatz* for sequent calculus. Finally, it should be mentioned that, for proving normalisation, one needs not only reduction functions, but also *permutations*, for which see (Prawitz, 1965).

[4]The way one conceives of atomic systems will normally have a significant impact on the final configuration of the Prawitzian semantics one works with. In particular, the account will be sensible to whether atomic rules are allowed to discharge assumptions, or to whether they are allowed to be *higher-order* rules, namely rules of level k having (and possibly discharging) rules of level $h < k$ as premises. For an overview see (Piecha, 2016).

semantically relevant for, as we shall see, atomic rules can be understood as a way for specifying the meaning of the non-logical terminology they concern, so that normalisation can be understood as a structural property of logical derivations which is invariant over this meaning determination.

Second, in many cases one can prove what (Schroder-Heister, 2006) has called the *fundamental corollary* of Prawitz's normalisation theory: normal derivations with no undischarged assumptions end by Gentzen's introductions. And this is semantically relevant for at least three reasons. First, derivations in sound systems can be reasonably understood as denoting *proofs*, so closed derivations can be reasonably understood as denoting *categorical* proofs, i.e. proofs which categorically establish their conclusions. At least since Frege, closed (saturated) entities are conceived of as prior to open (unsaturated) entities in one's semantics. E.g., that P holds of open entities is defined *by closure*, namely, by requiring P to hold of what one obtains by filling these entities with closed entities which enjoy P. Therefore, the corollary refers to the prior entities in a potential, proof-based semantics. Second, if derivations in sound systems denote proofs, they must be compelling with respect to the truth of their conclusions. And this must be because of some deductive kernel which compulsion stems from. If one accepts that detours play no actual deductive role, one easily sees that this deductive kernel is nothing but the normal forms which derivations reduce to. Third, proofs are convincing (also) because of the meaning of the formulas they involve. If these formulas are logically complex, the meaning of their logical constants must play an active role in the compelling character of proofs. And if, in the prior normal case, the derivations that denote these proofs end by introduction, a special connection must exist between meaning of the logical constants and Gentzen's introductions.

Observe that, above, I moved from speaking of derivations, which are *syntactic* objects in *formal* systems, to speaking of proofs, which are *semantic* in nature, since they are (thought of as) endowed with an epistemic force stemming from *meaning* (I leave aside the difficult question about whether a proof is an object or an act). In fact, the fundamental corollary seems to speak in favour, or even confirm, a famous *semantic* claim which natural deduction was born with, namely, Gentzen's well-known, proto-inferentialist idea that introductions fix the meaning of the logical constants while eliminations are unique functions of the corresponding introductions.[5] We may thus be

[5]That normalisation, a result essentially relative to formal systems, *confirms* a semantic claim, thereby triggering a normalisation-based semantics, may be read as an example of Kreiselian informal rigour. This view has been defended in (Montesi & Piccolomini d'Aragona, 2022).

entitled to generalise Gentzen's claim, and endorse Dummett's fundamental assumption, saying that, if *A* is categorically provable, then *A* is canonically provable, where "canonical" means that *A* is proved by introduction.

Note that *A* is required to be *provable*, not *derivable* in a system. Therefore, Dummett's assumption does not simply require eliminations *in some normalising system* to be unique functions of the corresponding introductions, but *arbitrary and generalised* eliminations to be – as one usually says – harmonic with respect to how meaning is fixed by introductions. This amounts to nothing but a semantic reading of Prawitz's *inversion principle* (Prawitz, 1965), which says that, if *A* is obtained by introduction, then its proof contains all the needed information for drawing consequences from *A*. When referred to formal systems, this principle concerns the possibility of devising reductions for the computation of detours, namely, sequences of introductions/eliminations *in that system*. But the principle can be also read semantically, i.e. as suggesting that valid inferences other than the introductions owe their deductive acceptability to a sort of stability with respect to how meaning is fixed by the introductions. Under this view, reductions for removing detours in formal systems are generalised towards *justifications* of arbitrary, non-introductory inferences. They show that these inferences preserve *validity* of arguments, rather than derivability of formulas in given formal setups.

3   From normalisation to proof-theoretic semantics

The question is now how the semantic insights which I have discussed in the previous section can be appropriately developed in such a way as to have a uniform semantic picture, a proper proof-theoretic semantics. In fact, I have already hinted at an answer to this issue – or better, at Prawitz's answer to this issue.

The two main ingredients of Prawitz's normalisation are: inferences and reductions. These must be generalised. Contrarily to derivations in formal systems, proofs do not employ inferences instantiating rules in some recursive set. For example, we do not think of our deductive activity as limited to some previously determined and fixed axiomatic framework. Thus, rather than speaking of derivations in formal systems, we speak of *argument structures*, i.e. sequences of formulas arranged in tree form in a Gentzenian natural deduction style. The nodes amount to *arbitrary* inferences, which may discharge assumptions. Given an argument structure, we say that it is *canonical* if it ends by an introduction, and that it is *non-canonical* otherwise, while we say that it is *closed* if it involves no undischarged assumptions, and that it is *open* otherwise.

Similarly, we have to generalise reduction functions and, as said, this should be done in such a way as to have transformation procedures which can be understood as *justifications* showing that arbitrary, non-canonical inferences preserve the validity of the arguments to which they are appended. For example, here is a transformation for disjunctive syllogism:

$$
\cfrac{\cfrac{\mathscr{D}_1}{\cfrac{A_h}{A_1 \vee A_2}} \, (\vee_I, i) \qquad \cfrac{\mathscr{D}_2}{\neg A_i}}{A_h} \qquad \text{reduces to} \qquad \cfrac{\mathscr{D}_1}{A_h}
$$

$(h, i = 1, 2, h \neq i)$. It is not difficult to realise that this procedure justifies disjunctive syllogism, i.e. that, if $\mathscr{D}_1$ and $\mathscr{D}_2$ are valid, then so is the output of the procedure – under the assumption that there can be no closed valid argument for $\bot$. More in general, we say that a *justification* for an inference R is an effective function ϕ defined on some sub-class K of the class of the argument structures ending by R and such that, for every $\mathscr{D} \in K$, $\phi(\mathscr{D})$ is an argument structure for the same conclusion from at most the same assumptions as $\mathscr{D}$. Given a set $\mathfrak{J}$ of justifications, an *expansion* of $\mathfrak{J}$ is a set of justifications $\mathfrak{J}^+$ such that $\mathfrak{J} \subseteq \mathfrak{J}^+$ and, for every $\phi \in \mathfrak{J}^+ - \mathfrak{J}$, the domain of ϕ is disjoint from the domain of any element of $\mathfrak{J}$ – justifications must be deterministic.[6]

Finally, an *argument* is a pair $\langle \mathscr{D}, \mathfrak{J} \rangle$, where $\mathscr{D}$ is an argument structure and $\mathfrak{J}$ is a set of justications.

4 Alternative definitions

We have now all the ingredients required for defining validity of arguments. Structurally, Prawitz's semantics does this in very much the same way as the one in which truth of formulas is dealt with in model-theory. In the latter, we

---

[6]We may renounce the latter restriction and define an expansion of $\mathfrak{J}$ as any set of justifications $\mathfrak{J}^+$ such that $\mathfrak{J} \subseteq \mathfrak{J}^+$. We would thereby obtain what (Schroder-Heister, 2006) calls *alternative justifications*. But here I will stick to (Prawitz, 1973). Let me finally remark that justifications ϕ are normally required to be linear over substitution, i.e., if A is an assumption of $\mathscr{D}$ and if we indicate with $\mathscr{D}^*/A$ the replacing of A with an argument structure $\mathscr{D}^*$ for A, then $(\phi(\mathscr{D}))[\mathscr{D}^*/A] = \phi(\mathscr{D}[\mathscr{D}^*/A])$.

first define truth of formulas in suitable set-theoretic structures, and then we generalise this to logical truth, understood as truth in all structures.

In Prawitz's semantics, set-theoretic structures are replaced by *atomic bases*, where an atomic base $\mathfrak{B}$ is a pair $\langle \mathfrak{L}, \Sigma \rangle$ with $\mathfrak{L}$ one's background language, and Σ an atomic system over $\mathfrak{L}$ (*expansions* $\mathfrak{B}^+$ of $\mathfrak{B}$ are defined in an obvious way). The idea is that the meaning of the non-logical terminology is not specified by mappings onto sets. Rather, the meaning of a non-logical sign k is given by stating the deductive behavior of k in atomic derivations, i.e. by giving rules which concern k in the atomic system. So for example

$$\overline{t + 0 = t} \qquad \overline{t + s(u) = s(t + u)} \qquad \overline{t \cdot s(u) = (t \cdot u) + t}$$

may be taken as fixing the meaning of + in Peano first-order arithmetic, as the usual binary function for addition on positive integers.[7] Thus, in Prawitz's semantics, we first define the notion of validity of an argument over a base, and then we define the notion of logical validity of an argument as validity of the argument over all bases.

However, because of the (normalisation-based) need of *both* evaluating the meaning of the non-logical signs and justifying non-canonical inferences, the situation in Prawitz's semantics is not as smooth as in model-theory. Already at the level of the ground notion of validity of arguments over bases, we have to choose between two alternative definitions. The crucial case is the one concerning validity of open arguments, since the by-closure definition may require arguments which replace assumptions to be valid on the very same base which we started from, or else on expansions of this base. To distinguish these two notions, in what follows I will call NE-validity the case when validity in the open case is defined without referring to expansions of the atomic base – thus, NE stands for *no expansions* – while I will call WE-validity the case when validity in the open case requires to take expansions of atomic bases into account – thus, WE stands for *with expansions*.

Definition 1 $\langle \mathcal{D}, \mathfrak{J} \rangle$ *is NE-valid over* $\mathfrak{B}$ *iff*

---

[7]Observe that the three rules above also involve the signs 0, s and $\cdot$, so the meaning of + is somehow connected with the meaning of those signs too. This requires stating some definitional order between atomic rules, for meaning determination to be sound (e.g. non-circular, or at least non-holistic, and so on). The idea that atomic bases determine the meanings of the non-logical signs is, I think, sufficiently clear from an intuitive point of view – and can be taken to be Wittgensteinian both in the sense that we tie meaning to use, and in the sense that use *shows* but does not *say* meaning. However, a more precise account, which may be compatible with Prawitz's semantics, can be found in (Cozzo, 1994) – where, e.g., one finds a precise definition of the notion of rules "concerning" symbols (in a direct or indirect way).

- $\mathscr{D}$ is closed $\Rightarrow$ $\mathscr{D}$ reduces through $\mathfrak{J}$ to a closed canonical argument structure which is valid over $\mathfrak{B}$ when paired with $\mathfrak{J}$;

- $\mathscr{D}$ is open $\Rightarrow$ if we replace its open assumptions with closed argument structures for these assumptions, which be valid over $\mathfrak{B}$ when paired with $\mathfrak{J}^+$, we obtain a closed argument structure which is valid over $\mathfrak{B}$ when paired with $\mathfrak{J}^+$.

Definition 2 $\langle \mathscr{D}, \mathfrak{J} \rangle$ is WE-valid over $\mathfrak{B}$ iff

- $\mathscr{D}$ is closed $\Rightarrow$ as in Definition 1;

- $\mathscr{D}$ is open $\Rightarrow$ for every $\mathfrak{B}^+$, if we replace its open assumptions with closed argument structures for these assumptions, which be valid over $\mathfrak{B}^+$ when paired with $\mathfrak{J}^+$, we obtain a closed argument structure which is valid over $\mathfrak{B}^+$ when paired with $\mathfrak{J}^+$.[8]

NE-validity is *non-monotonic*, in the sense that some argument is NE-valid on some $\mathfrak{B}$, without being NE-valid on some $\mathfrak{B}^+$. An example is the following: consider $\mathfrak{B} = \langle \mathscr{L}, \emptyset \rangle$, and consider the argument structure

$$\mathscr{D} = \frac{p}{q}$$

for p and q atomic. $\langle \mathscr{D}, \emptyset \rangle$ is NE-valid on $\mathfrak{B}$, simply because the antecedent in the definition of NE-validity in the open case is vacuously satisfied, i.e. there are no closed NE-valid arguments for p on $\mathfrak{B}$ – otherwise, they should reduce to closed derivations of p in the atomic system of $\mathfrak{B}$, which would contradict the emptiness of $\mathfrak{B}$. But consider now the expansion of $\mathfrak{B}$ given by $\mathfrak{B}^+ = \langle \mathscr{L}, \{A\} \rangle$, where A is the axiom

$$\frac{}{p} \, A$$

We have a closed argument for p which is NE-valid on $\mathfrak{B}^+$, i.e. A. But since we lack a closed argument for q which be NE-valid on $\mathfrak{B}^+$ – as q is not derivable in $\{A\}$ – the argument $\langle \mathscr{D}, \emptyset \rangle$ is not NE-valid on $\mathfrak{B}^+$ – it is non-canonical, and we have no information about how to reduce it. There *cannot* even be any set of justifications which makes $\mathscr{D}$ valid on $\mathfrak{B}^+$, because otherwise q should be derivable in $\{A\}$. Examples of this kind are blocked under the requirement that open-validity should propagate through expansions of the atomic base. In fact, WE-validity is *monotonic*.

---

[8]Of course, it is understood that closed derivations in the atomic system of $\mathfrak{B}$ are canonical and valid, and this is precisely because $\mathfrak{B}$ is thought of as fixing the meanings of the non-logical signs involved in such derivations, so we need no reduction.

Four Attitudes in Prawitzian Semantics

Theorem 1 (Schroeder-Heister, 2006) $\langle \mathcal{D}, \mathfrak{J} \rangle$ *is WE-valid over $\mathfrak{B}$ $\Leftrightarrow$ for every $\mathfrak{B}^+$, $\langle \mathcal{D}, \mathfrak{J} \rangle$ is WE-valid over $\mathfrak{B}^+$.*

Another result that we can prove with WE-validity is that logical validity is tantamount to WE-validity over the empty base $\mathfrak{B}_\emptyset$ – i.e. the base $\langle \mathfrak{L}, \emptyset \rangle$.

Theorem 2 (Schroeder-Heister, 2006) $\langle \mathcal{D}, \mathfrak{J} \rangle$ *is logically valid $\Leftrightarrow$ $\langle \mathcal{D}, \mathfrak{J} \rangle$ is WE-valid over $\mathfrak{B}_\emptyset$.*

Of course, Theorem 2 must be explained, since we have not yet defined what does it mean for an argument to be logically valid. Here, we have two further options, i.e. what I will call *system-rooted* validity – written SR – and what I will call *schematic* validity – written S.

Definition 3 $\mathcal{D}$ *is SR-valid iff, for every $\mathfrak{B}$ there is $\mathfrak{J}$ such that $\langle \mathcal{D}, \mathfrak{J} \rangle$ is valid over $\mathfrak{B}$.*

Definition 4 $\mathcal{D}$ *is S-valid iff there is $\mathfrak{J}$ such that, for every $\mathfrak{B}$, $\langle \mathcal{D}, \mathfrak{J} \rangle$ is valid over $\mathfrak{B}$.*

Observe that Definition 3 is best understood as a definition of logical validity, not for arguments, but for argument structures. The idea is that an argument structure can be consistently evaluated over every $\mathfrak{B}$ through a suitable $\mathfrak{B}$-depending $\mathfrak{J}$. This is much model-theoretic in spirit and indeed, Definition 3 can be turned into a notion of logical consequence for formulas which looks very much like the model-theoretic one (this is done in Definition 6 below, which requires Definition 5 first).

Definition 5 $\Gamma \models_\mathfrak{B} A$ *iff for some $\mathcal{D}$ for A from Γ, for some $\mathfrak{J}$, $\langle \mathcal{D}, \mathfrak{J} \rangle$ is valid over $\mathfrak{B}$.*

Definition 6 $\Gamma \models_{SR} A$ *iff, for every $\mathfrak{B}$, $\Gamma \models_\mathfrak{B} A$.*

However, this move seems not to be allowed with S-validity, i.e. we cannot reduce to Definition 5.

Definition 7 $\Gamma \models_S A$ *iff there is an S-valid $\mathcal{D}$ for A from Γ.*

Thus, Prawitz's semantics can be developed in two distinct ways at the local level and in two distinct ways at the global level. Is this completely co-incidental, or do the four options respond to some kind of inner equilibrium?

5   Conceptual duality: semantic *vs* deductive attitude

My point is that the alternatives *do* respond to an equilibrium, and indeed that they give rise to two groups of pairwise "symmetric" choices, where each pole of each group is constituted by a local and a global notion of validity – i.e. the group is a full-blooded proof-theoretic semantics. These symmetries are nothing but different attitudes one may soundly have towards Prawitzian constructivism, while the groups classify the attitudes under an either conceptual, or a structural duality. Let us start from the conceptual duality.

Definition 1 accounts for a quite natural idea about validity of arguments, i.e. that an argument may be valid mainly because of some special *meaning* of the non-logical terminology it involves. When we change this meaning, the argument may cease to be valid, just in the same way as, when we change the interpretation of a formula which happens to be model-theoretically true on some structure, the formula may get falsified. From this point of view, local validity should be non-monotonic, for rules may be acceptable only in virtue of some peculiar interaction with a given determination of non-logical meaning.

However, one may argue that when we speak of deductive correctness, we are referring to a property which is expected to remain stable over expansions of one's knowledge base. If something is established *deductively*, it must hold independently of any potentially new information. Therefore, although it is true that local validity may hold on some domains while failing on others, it should nonetheless be monotonic with respect to expansions of the atomic base. So, on this view, WE-validity is the right notion of local validity (recall that WE stands for *without expansions*, so it refers to Definition 2).

An advocate of NE-validity (recall that NE stands for *no expansions*, so it refers to Definition 1) may at this point observe that, in a sense, atomic bases could be well understood as knowledge bases, but this is not what they are primarily meant for. Rather, they should be principally understood as fixing the meaning of the non-logical signs.[9] Thus, if an argument remains valid throughout expansions of atomic bases, its validity does not really depend on non-logical meaning. When expanding the base, this meaning changes. Therefore, the strong point of WE-validity is a weak point in the perspective of NE-validity, and the strong point of NE-validity is a weak

---

[9]The opposition between these two ways of understanding atomic bases, and the implications of this opposition, are discussed in (Piecha, de Campos Sanz, & Schroder-Heister, 2015), but see also (Schroeder-Heister, 2018).

point in the perspective of WE-validity. For someone endorsing WE-validity, NE-validity is too focused on meaning and too little focused on deduction; for someone endorsing NE-validity, it is precisely the other way around. NE-validity provides a proof-theoretic *semantics*, while WE-validity provides a *proof-theoretic* semantics. We thus have two reasonable, although somewhat opposite attitudes one may have towards a proof-based semantics: a *semantic* attitude, and a *deductive* attitude.

Now, very much the same situation seems to occur, *mutatis mutandis*, at the global level. SR-validity (recall that SR stands for *system rooted*, so it refers to Definition 3) corresponds to the idea that an argument is logically valid when it is justifiable on every atomic base, namely, when the deductive framework built up by the order of its rules can be soundly adapted to every determination of non-logical meanings. The validity of the argument may be very much tied to this meaning determination, for nothing impedes that a justifications-set, while working on some bases, happens to fail on others – where the argument will be thus justified by a different set of justifications. Observe that, here, the semantic treatment of non-introductory inferences – i.e. the fact that their justification ranges over atomic bases – is similar to the treatment which *non-logical* signs undergo in model-theory, whose interpretation ranges over given structures.

Is this what we have in mind when speaking of logical validity of an argument? Do we really mean that the argument is *justifiable* on every base? Maybe not. Maybe we should deal with non-introductory inferences in the same way as model-theory deals with, not non-logical, but *logical* signs. We should require that the justification of the non-introductory inferences is fixed over the class of atomic bases, just like the interpretation of the logical signs remains fixed throughout model-theoretic structures. For, what we mean by logical validity of an argument is that the argument is *justified* whatever the base is, i.e. that we have a non-base-depending set of justifications. Thus, atomic bases, i.e. non-logical meanings, *play no actual role* as to whether the argument is correct, just like Theorem 2 required; we may content ourselves of operating locally on an empty base, for only the deductive structure of the argument matters. S-validity is the right notion (recall that S stands for *schematic*, so it refers to Definition 4).

To the eyes of an external observer, however, all the options may cope with this or that tenet of Prawitz's constructivism. So, the pairs $\langle NE, SR \rangle$ and $\langle WE, S \rangle$ constitute two coherent proof-theoretic semantics, of which the first one emphasizing the word 'semantics', and the second one emphasizing the word 'proof-theoretic'.

6 Structural duality: continuist *vs* discontinuist attitude

Should we conclude, from what said in the previous section, that the pairs $\langle \text{NE}, S \rangle$ and $\langle \text{WE}, SR \rangle$ do not amount to acceptable Prawitzian semantics? In fact, these readings are often found in the literature – see (Prawitz, 1973) for the former, and (Schroder-Heister, 2006) for the latter.

As I said at the beginning of Section 6, also in this case we can find a duality which the readings at issue respond to. However, at variance with the conceptual duality, this is not a duality about how we semantically conceive of proofs. Rather, this is a duality about how we conceive of the *structure* of our proof-based semantics. It concerns the way we conceive of the interplay between validity of arguments over atomic bases, and validity of arguments over all bases. Either we think of the latter as a simple generalisation of the former, based on the idea that what is logically valid is simply what is locally valid on each *locum*; or else we understand logical validity as validity in virtue of some special deductive force, whose impact in the local case is not as important as what renders an argument valid over a *locum*.

Since the structural duality concerns the overall articulation of our semantics, the best place for detecting it is the structure of our semantic definitions. The latter can be easily seen to be built up of four main "ingredients", applied to atomic bases and justifications sets. A *level*, either local or global. A *focus* element, that is, what the definition is about. Local validity mainly refers to bases, as we abstract from the justifications set, and just wonder whether validity in the open case should be defined relatively to the very same base which we started with, or else to expansions of it. Global validity, instead, mainly refers to justifications sets, as we quantify over – i.e. abstract from – atomic bases, and just wonder whether we should require justifications sets to be fixed, or else to range over bases. Accordingly, we also have an *abstraction* element. And finally, we have a focus *range*, i.e. an alternative about how to deal with the focus element. At the local level, we can choose whether validity in the open case is defined with respect to one and the same atomic base – the one which we started with – or else with respect to a class of atomic bases – the class of expansions of our base. Similarly, at the global level, we can choose whether logical validity is defined with respect to one and the same justifications set, working on all bases, or to a class of justifications sets, potentially one for each base. We can label our ingredients as follows:

- local level = L, global level = G;

Four Attitudes in Prawitzian Semantics

- focus on bases = $F^{\mathfrak{B}}$, abstraction on bases = $A^{\mathfrak{B}}$;

- focus on justifications = $F^{\mathfrak{J}}$, abstraction on justifications = $A^{\mathfrak{J}}$;

- one-and-the-same range = $O\&S$, class range = K.

Thus, the proof-theoretic semantics $\langle \text{NE}, SR \rangle$ (recall: NE stands for *no expansions*, i.e. Definition 1, while SR stands for *system-rooted*, i.e. Definition 3) can be described by the pair

$$\langle \langle L, F^{\mathfrak{B}}, A^{\mathfrak{J}}, O\&S \rangle, \langle G, A^{\mathfrak{B}}, F^{\mathfrak{J}}, K \rangle \rangle.$$

whereas the proof-theoretic semantics $\langle \text{WE}, S \rangle$ (recall: WE stands for *with expansions*, i.e. Definition 2, while SR stands for *schematic*, i.e. Definition 4) can be described by the pair

$$\langle \langle L, F^{\mathfrak{B}}, A^{\mathfrak{J}}, K \rangle, \langle G, A^{\mathfrak{B}}, F^{\mathfrak{J}}, O\&S \rangle \rangle.$$

The conceptual duality can be described at this point as an *inversion of the focus range*. The first pair has an $O\&S$ range in the first element, and a K range in the second element. The second pair goes the other way around, from a K range to an $O\&S$ range. The pairs also *invert the focus and abstraction elements*, as both focus on bases at the atomic level, and on justifications sets at the global level. Well then, but now we may wonder why we should not content ourselves with simply inverting the focus and abstraction elements, leaving the focus range unmodified. This would yield the pairs

$$\langle \langle L, F^{\mathfrak{B}}, A^{\mathfrak{J}}, O\&S \rangle, \langle G, A^{\mathfrak{B}}, F^{\mathfrak{J}}, O\&S \rangle \rangle$$

and

$$\langle \langle L, F^{\mathfrak{B}}, A^{\mathfrak{J}}, K \rangle, \langle G, A^{\mathfrak{B}}, F^{\mathfrak{J}}, K \rangle \rangle.$$

It is easy to see that this returns precisely the readings $\langle \text{NE}, S \rangle$ and $\langle \text{WE}, SR \rangle$ respectively.

I want now to argue that $\langle \text{WE}, SR \rangle$ constitutes a *continuist* attitude towards proof-theoretic semantics, in the sense that, if this approach is adopted, global validity can be understood as a generalisation of local validity. Similarly, $\langle \text{NE}, S \rangle$ amounts to a *discontinuit* attitude, for global validity is sensibly stronger than local validity. This can be seen from the following results.

Proposition 1 *In the implication-free fragment, it holds that $\langle \mathscr{D}, \mathfrak{J} \rangle$ WE-valid on $\mathfrak{B} \Rightarrow \langle \mathscr{D}, \mathfrak{J} \rangle$ NE-valid on $\mathfrak{B}$.*

Proposition 2 $\langle \mathscr{D}, \mathfrak{I} \rangle$ *S -valid* $\Rightarrow$ $\langle \mathscr{D}, \mathfrak{I} \rangle$ *S R-valid.*

In terms of focus ranges:

- by Proposition 1, at the local level $K \Rightarrow O\&S$ in the implication-free fragment;

- by Proposition 2, at the global level $O\&S \Rightarrow K$.

Proposition 1 may be understood as showing that WE-validity is stronger than NE-validity. The latter is non-monotonic everywhere and, at least in the $\{\wedge, \vee\}$-fragment of propositional logic, it is broader than the former; correspondigly, even in full propositional logic, NE-validity cannot imply WE-validity, but at most be implied by it, since we know that WE-validity *is* monotonic everywhere. If we accept this reconstruction, in $\langle WE, SR \rangle$ we move from something of *maximal strenght* locally, to something of *minimal strenght* globally. Of course, logical validity implies validity over a base, hence the former is stronger than the latter; but, if we look at the roles that our semantic notions play on the respective levels, we can realise that in $\langle WE, SR \rangle$ logical validity has *structurally* a lower global strenght than the local strenght of validity over a base. So, this approach can be understood as one where we simply take local validity, and strengthen it by quantifying over all bases; this continuity is not altered by the addition of any new ingredient.

In $\langle NE, S \rangle$ it is exactly the other way around. We move from something of *minimal strenght* at the local level, to something of *maximal strenght* at the global level. *Mutatis mutandis*, this produces a discontinuity. We do not simply take local validity and quantify over all bases; we add an ingredient which we had not at the local level, namely, the idea that logical deduction, being justified by one and the same set over all bases, owes its correctness to something more than variation of non-logical meaning.

7   Two tables and a diagram

Based on the classification proposed above of the ingredients of Prawitzian definitions of validity (focus element, abstraction element, O&S range and class range), the conceptual and structural duality can be more easily visualised through some schemes. The first table below shows the ingredients and kinds of validity. The second table describes the conceptual duality, which obtains by taking the diagonals (respectively, in braces and square brackets). Finally, the third table accounts for the structural duality, where it suffices to take the columns (respectively, in braces and square brackets).

	O&S range	Class range
Focus $\mathfrak{B}$, local level	NE	WE
Focus $\mathfrak{J}$, global level	$\models_S$	$\models_{SR}$

	O&S range	Class range
Focus $\mathfrak{B}$, local level	{NE}	[WE]
Focus $\mathfrak{J}$, global level	$[\models_S]$	$\{\models_{SR}\}$

	O&S range	Class range
Focus $\mathfrak{B}$, local level	{NE}	[WE]
Focus $\mathfrak{J}$, global level	$\{\models_S\}$	$[\models_{SR}]$

This shows that the dualities can be combined so as to give rise to the following diagram,

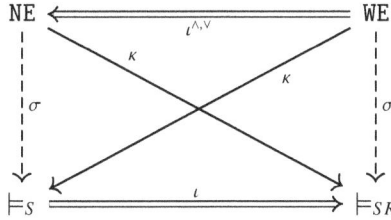

where κ indicates the conceptual duality, σ indicates the structural duality, and ι indicates the implications given by Propositions 1 and 2 (observe that the implication given by Proposition 1 is restricted to the implication-free fragment, whence the label).[10]

8   Conclusion

Let me now conclude with three remarks, possibly to be further dealt with in future works. First, as seen, both the conceptual and the structural duality

---

[10]In (Piccolomini d'Aragona, 2022b) I argue that any potential Prawitzian semantics can be obtained by taking pairs of 4-tuples out of the Cartesian product

$$\{L, G\} \times \{F^{\mathfrak{B}}, A^{\mathfrak{B}}\} \times \{F^{\mathfrak{J}}, A^{\mathfrak{J}}\} \times \{O\&S, K\}$$

where the first element of each 4-tuple in the pair is either L or G, and not both the first elements are L or G. I then prove that the diagram above is complete, in the sense that any potential Prawitzian semantics is either equivalent to one of the four we have already singled out, or semantically unacceptable. Thus the diagram provides an ordered and exhaustive classification of potential Prawitzian semantics.

consist of two poles which are, in a sense, symmetric to each other. The question may now be whether this happens also at the level of the dualities themselves, i.e. whether the dualities can be in turn understood as somehow "symmetric". One may e.g. observe that, by Propositions 1 and 2, the deductive attitude implies the semantic one at each level – bearing in mind that Proposition 1 is restricted to the implication-free fragment. Thus, the approaches belonging to the conceptual duality can be overall articulated into a sort of *linear* order, say $\langle \mathrm{WE}, S \rangle \Rightarrow \langle \mathrm{NE}, SR \rangle$. This does not hold in the structural duality for, as seen, here we have an inversion of the implicational strengths when moving from the local to the global level. Thus, while in the conceptual duality the local and the global notions are liable to a uniform treatment, in the structural duality this uniform treatment is not available.

Second, the dualities and attitudes I have highlighted stem from the interaction of non-logical meanings and justification of generalised eliminations. But one may think of a further element, namely, order of inferences, arguing that whether an argument is valid depends also on how inferences are arranged in the argument.[11]

The final question is the following. Can what I have been saying so far be applied, not only to Prawitz's semantics, but to proof-based semantics in general? Is the presence of this four, somewhat opposite, but all equally reasonable attitudes, not only a distinctive trait (indeed, a richness) of Prawitz's semantics, but also of semantic constructivsm as such? This would provide a quite sharp grid for classifying different types of constructivism, a sort of sequencing of its DNA.

References

Cozzo, C. (1994). *Meaning and argument. A theory of meaning centred on immediate argumental role.* Stockholm: Almqvist & Wiskell.

Francez, N. (2015). *Proof-theoretic semantics.* London: College Publications.

Gentzen, G. (1935). Untersuchungen über das logische Schließen i, ii. *Mathematische Zeitschrift, 39,* 176–210, 405–431.

Montesi, F., & Piccolomini d'Aragona, A. (2022). Prawitz's semantics and Walton's argument schemes: a tentative reading and application of Kreisel's informal rigour. *Noesis, 38,* 19-44.

---

[11]I am grateful to Ansten Klev for raising this point. It implies that the "proof-theoretic square" above could be turned into a "proof-theoretic solid".

Piccolomini d'Aragona, A. (2022a). Proofs, grounds and empty functions: epistemic compulsion in Prawitz's semantics. *Journal of philosophical logic*, *51*, 249-281.

Piccolomini d'Aragona, A. (2022b). The proof-theoretic square. Forthcoming.

Piecha, T. (2016). Completeness in proof-theoretic semantics. In T. Piecha & P. Schroeder-Heister (Eds.), *Advances in proof-theoretic semantics* (p. 231-251). Cham: Springer.

Piecha, T., de Campos Sanz, W., & Schroder-Heister, P. (2015). Failure of completeness in proof-theoretic semantics. *Journal of philosophical logic*, *44*, 321-335.

Prawitz, D. (1965). *Natural deduction. A proof-theoretical study*. Stockholm: Almqvist & Wiskell.

Prawitz, D. (1971). Ideas and results in proof theory. In J. E. Fenstad (Ed.), *Proceedings of the second Scandinavian logic symposium* (p. 235-307). Amsterdam: Elsevier.

Prawitz, D. (1973). Towards a foundation of a general proof-theory. In P. Suppes, L. Henkin, A. Joja, & G. C. Moisil (Eds.), *Proceedings of the Fourth International Congress for Logic, Methodology and Philosophy of Science, Bucharest, 1971* (p. 225-250). Amsterdam: Elsevier.

Prawitz, D. (2015). Explaining deductive inference. In H. Wansing (Ed.), *Dag Prawitz on proofs and meaning* (p. 65-100). Cham: Springer.

Schroder-Heister, P. (2006). Validity concepts in proof-theoretic semantics. *Synthese*, *148*, 525-571.

Schroeder-Heister, P. (2018). Proof-theoretic semantics. In E. N. Zalta (Ed.), *The Stanford Encyclopedia of Philosophy*.

Antonio Piccolomini d'Aragona
Institute of Philosophy, Czech Academy of Sciences
The Czech Republic
E-mail: `piccolomini@flu.cas.cz`

Theory of Predicables and Sortals

KAREL ŠEBELA[1]

Abstract: The paper aims to argue for the interpretation of Porphyry's theory of predicables in a system of modern logic, namely, sortal logic (SL). It is namely an attempt to interpret it adequately within the conceptual framework of modern logic, which is able to capture its essential features. Thus, the paper is focused on conceptual analysis and comparison. The theory of predicables (TP) is a classification of a certain type of predicates (first-order, monadic, simple). TP mainly establishes five different terms, genus, species, difference, property, and accident. The basic thesis of the interpretation in SL: let the species in TP be understood as a kind of sortals. The author interprets the concept of species as a complex sortal composed of a superordinate sortal concept and a specific difference concept, following in the footsteps of Aristotle. I believe that attempts to revive and reconstruct this doctrine could be useful and beneficial for current natural and philosophical language analysis, not to mention the history of logic, in which TP is of immense historical importance.

Keywords: sortals, Aristotle, predicables

1   Introduction

It is true that the very interpretation of Aristotelian logic (AL for short) with the help of one of the existing systems of modern logic is surely a nontrivial problem, following from the simple fact that AL is an ancient logical theory based on conceptions of logic, philosophy, and science (that are) relatively dissimilar to modern views. Thus, no choice is a trivial one; on the other hand, to better understand AL and aspire to extract something important from AL for contemporary/modern logic, some choices are desirable. The canonical interpretation of AL, which can be find in logical textbooks, take AL as a small part of the first-order predicate logic, as its mere fragment. This interpretation of AL has historical roots which are not the subject of this article.

---

[1] Work on this article was funded by the Czech Science Foundation as the project GA ČR 19-06839S Non-classical Interpretation of the Aristotelian Logic and Theory of Predication.

Karel Šebela

Importantly, this interpretation has some unwelcome consequences for AL. According to this interpretation, some key parts of AL simply do not work. Firstly, the square of opposition collapses—none of the relations within the square, except for contradiction, hold. Secondly, some of the syllogistic moods become invalid inference schemata (Darapti is the best-known case). Finally, the theory of categories and predicables becomes an extralogical matter.[2]

Therefore, what if there is another logical system, in which the interpretation of AL would be more felicitous? In this article, I would like to offer reasons for a conservative modification of classical first-order logic, namely sortal logic, as an alternative. This logic will be introduced in the following section. However, as a motivation, I would like to mention some results achieved by sortal logic in the field of interpreting AL. As it was demonstrated, e.g., in (Smiley, 1962), in sortally interpreted AL, the square of opposition works correctly and all syllogistic moods (including Darapti) hold. Thus, a lot of work has already been done. What remains unclear is the role and the final systematic conception of categories and predicables in AL. The main purpose of this article is to start to address this point.

Why deal with the theory of predicables (TP) today? The theory of predicables can be seen as a classification of a certain class of predicates, namely unary first-order predicates. A similar classification is missing in modern logic; and as these predicates are used very often when we speak about things around us, it seems preferable to have a classification of such predicates at the very least. Moreover, as a traditional part of logic, it is closely linked to and used in philosophy. The distinction between different types of predicates seems to be supported by natural language and argumentation (terms like species, property, etc., are common even in everyday usage, apart from Linné's classification). It is also very common to treat concepts as subordinated and superordinated. The difference between essential and non-essential predication is also of great philosophical importance. Therefore, I believe that attempts to revive and reconstruct this doctrine could be useful and beneficial for current natural and philosophical language analysis, not to mention the history of logic, in which TP is of immense historical importance. The theory of predicables is a traditional part of logic and we can find multiple versions of it. In this paper, we will limit ourselves to Porphyry's version of the theory as it is outlined in his *Isagoge*. I am fully

---

[2]For details, see (Parsons, 2021).

154

aware that Porphyry's version may differ substantially from later scholastic versions. We will use the methods of conceptual analysis and comparison.

In short, the aim of the article is to argue for the interpretation of Porphyry's theory of predicables in a system of modern logic, namely, sortal logic (SL).

2   Predicables

The theory of predicables (TP) is a classification of a certain type of predicates (first-order, monadic, simple). TP mainly establishes five different terms: genus, species, difference, property, and accident. Furthermore, a traditional distinction is made between essential and nonessential predicables. It is a question of whether a given predicable is the answer to the question, "What is it?" If so, it is an essential predicable, if not, it answers the question, "What is it like?" Essential predicables include genus and species, and nonessential ones include accident, property, and difference. Here we can find one of the greatest divergences between Porphyry's theory of predicables and the scholastic version, where the notion of difference is an essential predicable. But in Porphyry we can find explicitly that difference answers the question "What is it like?". This nonessential interpretation of difference will be commented on below.

An important part of Porphyry's theory of predicables is so-called Porphyrian tree (Porphyry, 2003, p. 5–7). It is an ordering of genera and species within the category of substance. The tree shows the subordinated relations between a set of predicates and moreover the role of differences in dividing genera into subordinated species.

3   Sortal logic

What is a sortal logic? For the sake of simplicity, it can be understood through a basic intuition in this example: in the context of geometry, consider the sentence "x lies on y." It seems to be natural to say that in this case, x ranges over points, whereas y ranges over lines. Thus, here the variables range over different kinds of objects. Nonetheless, in classical first-order logic, all variables range over one general domain of objects.

Thus, sortal logic (SL for short) is an attempt to build up a logical system based only on various kinds of objects from one domain. Generally, logic is sortal when it includes a set of individual variables and these variables

Karel Šebela

are bounded by sortal quantifiers, which explicitly involve a sortal predicate. This theory is a conservative enlargement of classical first-order predicate logic. First-order logic can therefore be called one-sorted logic. The very term "sortal" denotes a predicate, which can only be predicated (assigned to/applied to) about the given kind of objects. The term is therefore a class constant, e.g., a sort of men, which is the extension of the predicate "man."

Sortal quantification subsequently looks as follows: instead of one-sorted quantification $(\forall x)$, the parentheses also indicate a sortal, which limits the quantification, so if we designate a sortal "man" with the symbol "S", then the quantification "for every man ..." is symbolised as $(\forall xS)$. Rewriting the traditional Aristotelian division of judgements into the SL looks like the following:

- $SaP - (\forall xS)Px$

- $SeP - (\forall xS)\neg Px$

- $SiP - (\exists xS)Px$

- $SoP - (\exists xS)\neg Px$

If one would like to introduce in SL the "classical" universal and existential quantification, then it could be done (in the second-order language) in the following way:

- $(\forall x)\varphi =_{\mathrm{df}} (\forall S)(\forall xS)\varphi$

- $(\exists x)\varphi =_{\mathrm{df}} (\exists S)(\exists xS)\varphi$ [3]

Regarding categories, the debate about the concept of a sortal is more important here because a sortal offers itself as a modern counterpart of the concept of category. We will not keep track of the whole debate; for the purposes of this article, only those basic characteristics are important, and they are commonly agreed upon. Firstly, a sortal is a predicate that enables the classification and counting of instances of that predicate.

In other words, it receives numerals as modifiers. A more philosophical (conceptualistic) definition is given by Nino Cocchiarella: It is "a socio-genetically developed cognitive ability or capacity to distinguish, count, and collect or classify things."[4]

[3](Cocchiarella, 1977, p. 443).
[4](Cocchiarella, 1977, p. 441).

In some versions of SL, sortals are subordinated to other sortals, so we can build a hierarchy of sortals. Moreover, in these conceptions, there are ultimate, i.e. non-subordinate sortals, i.e., sortals that are not subordinated to any other sortal. These versions of SL seem to be the closest to the Aristotelian theory of categories, with its hierarchy of genera and species and with the Porphyrian tree.[5] Therefore, we will use these versions.

4   Why sortal logic?

First-order logical systems (classical or otherwise) usually do not distinguish within the group of predicates. For SL (and for the goals of this article), there is quite a substantial difference between the two kinds of predicates that, for simplicity, can be labelled as sortal and non-sortal predicates.

Generally speaking, while sortal predicates include many of the countable common nouns, like "cat" and "house", non-sortal predicates include most adjectives, intransitive verbs, and above all, mass terms, e.g., "knowledge," "cheese," or "water." Along with this distinction, we can distinguish (between) sortal and standard monadic predications. A sortal predication is then simply a predication with sortal predicates, such as the one in the statement "John is a person" or generally in sentences of the form "a is an S" (where S is a sortal predicate). Standard predication is a predication with non-sortal predicates, such as "John is German" and "John is running," or generally in sentences of the form "a is P" (where P is a non-sortal predicate). Needless to say, the distinction between sortal and non-sortal predicates is very similar to the distinction between essential and non-essential predicates and the distinction between sortal and standard predication is very similar to the distinction between essential and non-essential predication. The reason for this similarity lies in the similarity between the very notions of a sortal and essential predicate. Both kinds of predicates serve *inter alia* to classify and identify kinds of things, and both of them have a special role in characterising what a given thing is. Of course, there are some dissimilarities, not every sortal is an essential predicate, e.g., sortals like architect, house. The reason is rather metaphysical, in the case of houses it is argued they do not have essences, they are mere aggregates. I would like to stay just on the level of conceptual analysis, so I will leave these metaphysical debates aside. The case of architect (and similar cases like philosopher or woman) exemplifies a mixed case, which is partly essential and partly accidental. These predicates

---

[5] See (Freund, 2019).

are also sortals, but not essential predicates. More problematic is the case of concepts expressed by mass terms, like water, which are essential predicates, but not sortals. The semantics of mass terms is a notoriously complicated issue and I have to admit that these terms constitute an exemption from this similarity. In sum, the similarity between sortals and essential predicates comprises all essential predicates except mass terms and for the sake of metaphysical controversies, we left aside also mixed cases and aggregates.

5   Species, genus in SL

According to Porphyry's *Isagoge*, a genus is that under which a species is ordered, and species is what is ordered under a genus. It means species and genera are correlative concepts, e.g., animal is a genus relatively to man, because man is a species of animal. At the same time, animal is a species of body. Animal is therefore a genus with respect to man and a species with respect to body. This is true for every species and genus, except for the lowest species, i.e., species that no longer divide into other species (such as humans), and the highest genera, i.e., genera that are no longer species of some superior genus (such as categories).

This opens the door to the following thesis, which is the basic thesis of Aristotelian sortal logic (TASL); let the species in Aristotelian logic be understood as a kind of sortals. It definitely does not hold the other way around, i.e., that every sortal would be a species. As mentioned in section 4, we have to leave aside the mass terms. The other problematic case is accidental species, like "courage", "the length of 1 meter", or "similarity". The problem is that these predicates are *prima facie* not sortals, but they are species. The problem seems to be on the ontological level because what falls, e.g., under a concept of courage are individual courages, i.e., accidents of individual men. In SL, equally as in the standard first-order logic, what falls under a concept, is an object, an individual. So, in order to include the accidental species, it will be necessary to enrich our ontology to contain not only objects but also even accidents of these objects. This enrichment is possible but in this article we will limit ourselves only to the ontology of objects.

TASL is therefore formulated for species, but it means the correlative concept of species. Furthermore, we will see that the definition of a species in ASL does not apply to the highest genera, although it does apply to the lowest species. In what follows, whenever we will speak about species, we will mean it as a common term for lowest species, species, and genera, with the exception of the highest genera.

As was mentioned in the section about SL, sortals are taken as connected with a principle of identity, i.e., according to Strawson (1959, p. 186), "a principle for distinguishing and counting individual particulars." According to the Aristotelian definition of a species, each species is a compound of the nearest higher genus and specific difference. TASL refers to the concept of species understood in this way. Purely for simplicity of expression, I will introduce a convention where I will denote the concept of species understood in this way as $\Phi(\alpha, \beta)$; Φ is a given species (e.g., man), α is a superordinate sortal (genus proximum, e.g., animal), and β is a specific difference (e.g., rational). Therefore, β is not a sortal, it is rather an adjectival term in Geach's (1962) terminology. Thus, in sortal Aristotelian logic (SAL for short), we accept the idea of superordinate sortals. The possibility to build a hierarchy of sortals is one of the things which connects SL closely to AL. The well-known Porphyrian tree can thus be seen as a good example of a hierarchy of sortals.

More specifically, we can make use of the syntax of the logic of complex sortal concepts, as formulated in Freund (2019), in which complex sortals are expressed in terms of the lambda operator. The idea is that complex predicates cannot be easily treated as representing sortal terms and Church's lambda operator can fix that. What we shall have then are lambda abstracts of the form $(\lambda x S.\varphi)$, where φ is well-formed formula of the language and S a sortal term. The lambda operator is sortally restricted, which means that individual variables bounded by the lambda operator will have to be within the scope of a sortal term. E.g., a complex predicate *car that is red* can be represented as $(\lambda x\, Car.Red(x))$. In our case, we can have lambda abstracts of the form $(\lambda x\alpha.\beta)$. One can spell out the lambda expression in informal terms as being an α such that β.

Now, returning to the α and β distinction, Dummett (1973) suggests that sortal logic is also connected with a criterion of application which determines when it is correct to apply a predicate to an individual. Most adjectives, according to Dummett, are connected only with a criterion of application. So β shares the same criterion of application as "its" α. As Dummett writes:

> To grasp the sense of either kind of general term, we have to learn the associated criterion of application under what conditions it is true to say of an object that it is a man, or that it is red. However, with general terms of the former kind, we have also

learnt something else-the associated criterion of identity-which is not, or at least not completely, determined by the criterion of application: we have, e. g., to learn what '. . . is the same man as . . .' means. With general terms of the second kind, however, there is nothing more to learn: the expression '. . . is the same red thing as . . .' has no univocal sense, and can be supplied with one only by giving the word 'thing' some specific content.[6]

Therefore, specific differences are always genus-relative, i.e., they are meaningfully applied only to the members of the given superordinated sortal (= genus). Now, when we predicate Φ of some object a, then we claim a) that it falls under the sortal a (and thus it determines the criteria of the identity of a) and b) that we determine the object thus identified in more detail using β. In this way a) a is a case of sortal predication, and b) is a case of standard predication.

6 Difference in SL

Now it will be suitable to try to specify β, that is, which type of adjective it is. Importantly, a difference has the ability to divide its genus.

The union of β and the negation of β give the extension of the genus. It could be objected that more adjectives not listed in the index of differences can have this property, e.g., it seemingly holds for all animals that they are black or non-black. Thus, one must add another specification. In accordance with Porphyry, we could distinguish differences from other adjectives (properties and accidents in Porphyry's vocabulary) by the fact that they have the ability to divide their genus, they cannot stand for only a part of the species (and in most cases, they stand for more than one species), and species are not predicated about them. But the main specification of the difference is the role it can play in a kind of complex sortals, namely species. So the question of whether a given adjective is a difference is the question of whether a given adjective plays this role. It could be objected that traditionally in Aristotelian logic a difference is taken as an essential predicate and also in the present conception it is a part of essential predicate. The argument goes as follows: if a species is a (purely) essential predicate, then any predicate by means of which a species is defined must also be essential. But the difference is part of the definition of a species. But I do not think that any predicate by means

---

[6](Dummett, 1973, pp. 547–548)

of which a species is defined must also be essential. What makes a predicate essential is the specific combination of predicates. It is let's say a new quality which emerges from the combination. And a very fact that a given predicate is essential or not (and so that it is of a given structure or not) is not a matter of conceptual analysis, but of an investigation of things to which a given predicate applies. But I admit the persuasiveness of the above argument and I do not want to interfere with the scholastic debates about the (non)essentiality of difference. In this text I would like to focus on the Porphyry's version and take the essenciality vs. nonessenciality division as simply mirroring the linguistic division between substantives and adjectives. The question whether this is a purely linguistic difference (as Leibniz believed) and to what extent it is derived from the structure of reality itself, the whole subsequent tradition has debated at length - but Porphyry himself wants to avoid these ontological questions and so do I. In the present interpretation, what distinguishes a difference from other adjectives is precisely that it occurs as part of a complex predicate of a species.

Finally, we must think about the negation of a difference. It is widespread for Aristotelian logic to speak about the nonrational, non-animate (inanimate), incorporeal, etc. In the Porphyrian tree, these negative differences from the right side of the tree, seemingly with the same level of legitimacy as the left side of the tree. It seems that the negation of the difference is a case of internal negation because here the scope of the negation is limited to the appropriate genus. The difference "nonrational" clearly applies to animals only, not to plants or minerals. However, keep in mind that the difference is always relative to a given genus, which in sortal logic can be expressed by sortal quantification. Thus, the statement that every animal is rational or non-rational can be expressed by $(\forall x S)(R(x) \vee \neg R(x)))$. Here, the second disjunct clearly denotes only non-rational animals, not simply everything that is not rational. However, here the limitation is given by the sortal quantification, so the negation $\neg$ could be understood as a classical Boolean negation.

7 Property and accident

Traditionally, a property is a predicate that does not indicate the essence of a thing, yet it belongs to that thing alone, and it is predicated convertibly of it.

Accidents are, according to Porphyry, items that come and go without the destruction of their subjects. But the characteristics fit better with the accidents as real parts of things. Accidents as a special type of predicate are predicates, that hold of their subjects contingently.

Thus, property and accident are just as difference non-sortal predicates, i.e., they are always genus-relative, and so their predication is a case of standard predication.

A special feature of properties is their coextensionality with the species. At the intensional level, properties can be taken as follows from the concept of the respective species. A traditional example could be the ability to laugh and rationality. Rationality is the specific difference of man, and in this sense, it is a part of the concept of man. The ability to laugh is not part of the concept but it is traditionally taken as a consequence of rationality.

Accident then differs from property precisely in that it is predicated only about a part of the extension, or in the temporal version, only about certain time phases of the same individual; in the modal version, it is not necessary for an object to exemplify it.

8   Conclusion

I tried to argue that the concept of predicables from Aristotelian logic can be successfully reconstructed in modern sortal logic. The key thesis is that species in Aristotelian logic be understood as (complex) sortals.

The criterion of identity of sortals conceived in this way makes it possible to clarify the special place of the highest genera in the hierarchy of categories. The highest genus cannot be conceived of as a compound of the form $\Phi(\alpha, \beta)$, because in this case, there is no α. Therefore, the highest genera either are already quite simple concepts, without internal structure, or their structure needs to be grasped in a different way than the one I tried to sketch in this article.[7]

References

Cocchiarella, N. (1977). Sortals, natural kinds and re-dentification. *Logique et Analyse*, *20*, 438–474.
Dummett, M. (1973). *Frege: Philosophy of Language*. London: Duckworth.
Freund, M. (2019). *The Logic of Sortals*. Cham: Springer.

---

[7]The author thanks the anonymous reviewers for their careful reading of the manuscript and their many insightful comments and suggestions.

Parsons, T. (2021). The traditional square of opposition. In E. N. Zalta (Ed.), *The Stanford Encyclopedia of Philosophy* (Fall 2021 ed.).

Porphyry. (2003). *Introduction*. Oxford: Oxford University Press. (Translated, with a commentary, by Jonathan Barnes)

Smiley, T. (1962). Syllogism and quantification. *Journal of Symbolic Logic*, *27*, 58–72.

Strawson, P. F. (1959). *Individuals: An Essay in Descriptive Metaphysics*. London: Methuen.

Karel Šebela
Palacký University in Olomouc, Faculty of Arts
Czech Republic
E-mail: karel.sebela@upol.cz

Remarks on Semantic Information and Logic. From Semantic Tetralateralism to the Pentalattice 65536_5

HEINRICH WANSING[1]

Abstract: A 16-element lattice 16_{inf} of generalized semantical values pre-ordered by set-inclusion as an information order is presented. The propositional logic Inf of that lattice is axiomatized and a generalization of 16_{inf} to a 65536-element pentalattice is suggested.

Keywords: Informational interpretation; Kripke semantics; Semantic tetra-lateralism; Generalized semantical values; Negation inconsistency.

1 Introduction

The paper deals with the notion of semantic information carried (or conveyed) by a declarative sentence, especially information carried by a formula in certain propositional languages in a given model in virtue of the meaning of the logical operations. The focus is thus on *logical information* and not on information in terms of the descriptive content of internally structured atomic sentences in first- or higher-order languages. If the information carried by a formula A in a model is represented by sets of states at which A is semantically evaluated, then 'classically' the evaluation gives rise to a distinction between two sets, the set of states at which A is true, A's truth set in the model, and the set of states at which A is false. The information carried by A is given already with A's truth set (also called 'the UCLA proposition expressed by A'), because falsity is identified with untruth and A's truth set determines its complement as A's falsity set. If we shift our attention

---

[1] I am grateful to Satoru Niki, Yaroslav Shramko, and an anonymous referee for helpful comments on a draft version of this paper. This research has received funding from the European Research Council (ERC) under the European Union's Horizon 2020 research and innovation programme, grant agreement ERC-2020-ADG, 101018280, ConLog.

from truth and falsity to information given with respect to the truth or falsity of atomic formulas and ultimately arbitrary formulas, we are dealing with what Nuel Belnap (1976, 1977) has called 'told values': **T** (*told true but not false*), **F** (*told false but not true*), **N** (*told neither true nor false*), **B** (*told both true and false*). The information carried by a formula A in a model is then represented by four sets of states, and the set of states at which a formula A is told false need not coincide with the set of states at which A fails to be told true.

The states of a model can be seen as *information states* as they represent the semantic information that is given with a valuation function. With Belnap's four-valued functions, a state may support the truth or the falsity of an atomic formula, and if no combination of being told is excluded, there may be states at which a given atomic formula is both told true and told false (states that support both the truth and the falsity of the formula) and states at which the formula is neither being told true nor being told false (states that neither support the truth nor the falsity of the formula). As is well known, the set of states can be given a relational or algebraic structure. In Grzegorczyk's (1964) and Kripke's (1965) informational interpretation of intuitionistic logic, the non-empty set of states is pre-ordered or partially ordered by a binary relation of possible expansion of information states. The semantics is made many-valued in the relational semantics for Nelson's constructive logics with strong negation **N3** and **N4**, see (Odintsov, 2008) and references therein, by introducing two separate satisfiability relations, verification (support of truth) and falsification (support of falsity). Informationally interpreted algebraic structures for substructural subsystems of intuitionistic logic and Nelson's logics, namely models based on semilattice-ordered monoids, have been studied in (Wansing, 1993a; 1993b). Also in Urquhart's (1972) semilattice semantics for relevance logic the set of states has an algebraic structure, featuring a binary operation of combination of information states (or pieces), see also (Punčochář, 2016), (Weiss, 2022). The ternary relation used in Routley-Meyer models for relevance logic has been given an informational reading by Mares (2009, 2010) and, more recently, Punčochář and Sedlár have developed an *information based semantics* in the context of inquisitive logic (Punčochář, 2019), (Punčochář & Sedlár, 2021).

Whilst the use of such relational and algebraic information structures turned out to be a rich and flexible approach in the study of substructural and other non-classical logics, I will focus on further semantical categories in addition to truth and falsity, respectively support of truth and support of falsity. With the distinction between sense and reference, Gottlob Frege enriched the inventory of basic semantical categories and values. Next to truth and falsity there are meaningfulness and meaninglessness (nonsensicality). Although according to Frege in a scientific language it ought to be the case that the sense of a sentence (the thought expressed by it) determines the sentence's reference (its truth value *The True* or *The False*), Frege nevertheless acknowledged natural language sentences that have a meaning but no reference. The four basic semantic values (*true, false, meaningful,* and *nonsensical*) induce a set of sixteen told values, including the values *told both meaningful and false* and *told both meaningful and nonsensical*. In this paper I will present two non-classical logics in languages that contain the unary connectives $[m]$ ("it is meaningful that") and $[n]$ ("it is nonsensical that"). One system, **N4mn**, is an expansion of the four-valued constructive and paraconsistent logic **N4**, and it is presented in (Wansing & Ayhan, 2023) as a case study in logical tetralateralism.[2] The other system, $\mathbb{Inf}$, is a logic interpreted on a 16-element lattice $\mathbb{16}_{inf} = (\mathbb{16}, \subseteq)$ of generalized truth values generated from the set of the four basic semantical values by considering its powerset, $\mathbb{16}$. In **N4mn** (and its connexive version **C4mn** defined in Section 5), the information carried by a formula A in a model is represented by 16 sets of states, in $\mathbb{Inf}$ it is represented by one out of 16 semantical values.

The move from metaphysically understood semantical values to informational told values allows one to take a fresh look at logical consequence and hence on logic. On the standard conception, semantic consequence is understood as truth preservation from the premises to the conclusion of an inference, and, from a 'classical' point of view, as untruth preservation from the conclusion to the premises. From the informational point of view, one may think of logic as the study of information flow, see (Mares, 2008), (Wansing, 2022), (Wansing & Odintsov, 2016).[3] Information flow, however, comes in more than one flavor depending on the basic semantic categories. In a valid inference, the information that the premises are true, false, meaningful, respectively nonsensical provides the information that the conclusion is true, false, meaningful, respectively nonsensical; that is, if the premises are told true, false, meaningful, respectively nonsensical, then so is the conclusion.

In the paper, the 16-valued logic **N4mn** is introduced semantically and shown to be faithfully embeddable into positive intuitionistic propositional

---

[2]The term 'tetralateral' mixes Greek and Latin. Such a mixture is, however, not unusual and can also be found, for example, in the expressions 'tetra-lateral position sensing detectors' and 'tetravalued modal algebras'.

[3]An informational account of entailment in terms of informational content inclusion has been suggested in (Shramko & Wansing, 2021).

logic. The logic $\mathbb{Inf}$ is new. It is introduced as a formula-formula inference system and is shown to be sound and complete with respect to $\mathbb{16}_{inf}$. Semantic consequence is defined with respect to the subset relation as an information order on $\mathbb{16}$, and set intersection (union) as the lattice meet (join) gives rise to a conjunction (disjunction) connective. The presentation ends with the definition of a 65536-element pentalattice, $\mathbb{65536}_5$, with five lattice orderings: an information preorder, a truth preorder, a falsity preorder, a meaningfulness preorder, and a nonsensicality preorder. This step is motivated by the rationale for proceeding from the smallest non-trivial bilattice $FOUR_2$ to the trilattice $SIXTEEN_3$, see Shramko and Wansing (2005, 2011).

2 Meaning and information

In this section I will address some basic terminological and conceptual issues.

The word 'information' is used in different ways in different contexts. Nevertheless, as Luciano Floridi (2010, p. 20 f.) explains:

> Over the past decades, it has become common to adopt a General Definition of Information (GDI) in terms of data + meaning. GDI has become an operational standard, especially in fields that treat data and information as reified entities, that is, stuff that can be manipulated (consider, for example, the now common expressions 'data mining' and 'information management'). A straightforward way of formulating GDI is as a tripartite definition (Table 1): According to (GDI.l), information is made of data. In (GDI.2), 'well formed' means that the data are rightly put together,

Table 1. The General Definition of Information (GDI)

GDI)	σ is an instance of information, understood as semantic content, if and only if:
	GDI.l) σ consists of *n data*, for $n \geq 1$;
	GDI.2) the data are well formed;
	GDI.3) the well-formed data are *meaningful*.

> according to the rules (*syntax*) that govern the chosen system, code, or language being used. ... Regarding (GDI.3), this is where semantics finally occurs. 'Meaningful' means that the data must comply with the meanings (*semantics*) of the chosen system, code, or language in question.

Remarks on Semantic Information and Logic

Given the looseness of the term 'information', the GDI is a solid basis
to work with. If the data one is interested in are declarative sentences from
a natural language or formulas from a formal language (closed formulas in
the case of a first- or higher-order language), the data are well-formed, and
the information carried or conveyed by a declarative sentence or formula, its
semantic content, is its meaning. A meaningful compound sentence (formula)
consists of subsentences (subformulas), each of which is well-formed and,
moreover, meaningful if we assume compositionality of meaning.

There is more to be said about the concept of semantic information, but
in what follows by 'semantic information' I will understand the meaning of
a declarative sentence and, in particular, the meaning of a formula from a
given formal language.

3 Semantic tetralateralism

Preparatory to the introduction of the logic 𝕀𝕟𝕗 in Section 4, we will expand
the language of propositional **N4** by two unary connectives, $[m]$ and $[n]$. A
formula $[m]A$ is to be read as "it is meaningful that A", and $[n]A$ is to be
understood as "it is nonsensical that A". The logic of the expanded language
will be referred to as **N4mn**. Its semantics is a tetralateralism insofar as it
makes use of four different forcing relations.

The propositional language $\mathcal{L}$ of **N4mn** based on a denumerable set of
propositional variables Φ is defined in Backus-Naur form as follows:

variables Φ: $p \in \Phi$
  formulas: $A \in Form_{\mathcal{L}}(\Phi)$
    $A ::= \quad p \mid (A \wedge A) \mid (A \vee A) \mid (A \rightarrow A) \mid \sim A \mid [m]A \mid [n]A.$

The language $\mathcal{L}'$ of positive intuitionistic propositional logic, **IPL**$^+$, is
obtained from $\mathcal{L}$ by dropping the unary connectives, i.e., $\sim$, $[m]$, and $[n]$,
and the language $\mathcal{L}''$ of the propositional logic **N4** is obtained from $\mathcal{L}$ by
dropping $[m]$ and $[n]$.

Definition 1 *A Kripke frame is a structure $\langle M, R \rangle$, where M is a nonempty
set (of information states), and R is a reflexive and transitive binary relation
(of information state expansion) on M.*

Definition 2 *A valuation $\models$ on a Kripke frame $\langle M, R \rangle$ is a mapping from
the set Φ of propositional variables to the power set 2^M of M such that for
any $p \in \Phi$ and any $x, y \in M$, if $x \in \models (p)$ and xRy, then $y \in \models (p)$. We*

169

will write $x \models p$ for $x \in \models (p)$. A valuation $\models$ is extended to a mapping from the set of all $\mathcal{L}'$-formulas to 2^M by:

$$x \models A{\rightarrow}B \text{ iff } \forall y \in M \, [xRy \text{ and } y \models A \text{ imply } y \models B],$$
$$x \models A \wedge B \text{ iff } x \models A \text{ and } x \models B,$$
$$x \models A \vee B \text{ iff } x \models A \text{ or } x \models B.$$

If $\mathcal{F} = \langle M, R \rangle$ is a Kripke frame, then $\langle M, R, \models \rangle$ is a Kripke model for **IPL**$^+$ *based on $\mathcal{F}$.*

The following *heredity condition* holds for $\models$: for any $\mathcal{L}'$-formula A and any $x, y \in M$, if $x \models A$ and xRy, then $y \models A$.

Definition 3 *An $\mathcal{L}'$-formula A is* true *in a Kripke model $\langle M, R, \models \rangle$ for* **IPL**$^+$ *if $x \models A$ for any $x \in M$, and is* valid *on a Kripke frame $\mathcal{F} = \langle M, R \rangle$ if it is true for every Kripke model for* **IPL**$^+$ *based on $\mathcal{F}$. An $\mathcal{L}'$-formula A is said to be* **IPL**$^+$-*valid if A is valid on every Kripke frame. Let $\Gamma \cup \{A\}$ be a set of $\mathcal{L}'$-formulas. Semantic consequence (entailment) is defined in terms of truth preservation at each state: $\Gamma \models A$ if for every Kripke model $\langle M, R, \models \rangle$ for* **IPL**$^+$ *and for all $x \in M$, $x \models A$ if $x \models B$ for all $B \in \Gamma$. We define the logic* **IPL**$^+$ *model-theoretically as the pair $\langle \mathcal{L}', \{\langle \Gamma, A \rangle \mid \Gamma \models A\} \rangle$.*

We turn to the language $\mathcal{L}$ and define *four* separate valuation functions $\models^+, \models^-, \models^m,$ and $\models^n$. These mappings determine for a given propositional variable p the set of states that support the truth, the falsity, the meaningfulness, and the nonsensicality (meaninglessness) of p, respectively. Support of truth, support of falsity, support of meaningfulness, and support of meaninglessness are seen as properties that are independent of each other. In particular, it is not excluded that an information state supports both the truth and the falsity of a given propositional variable or both its meaningfulness and its nonsensicality.

Definition 4 *The valuation functions $\models^+, \models^-, \models^m,$ and $\models^n$ on a Kripke frame $\langle M, R \rangle$ are mappings from the set Φ to the power set 2^M of M such that for any $\star \in \{+, -, m, n\}$, any $p \in \Phi$ and any $x, y \in M$, if $x \in \models^\star (p)$ and xRy, then $y \in \models^\star (p)$. We will write $x \models^\star p$ for $x \in \models^\star (p)$. The functions $\models^+, \models^-, \models^m,$ and $\models^n$ are extended to mappings from the set of all $\mathcal{L}$-formulas to 2^M by:*

$$x \models^+ A \wedge B \text{ iff } x \models^+ A \text{ and } x \models^+ B,$$
$$x \models^+ A \vee B \text{ iff } x \models^+ A \text{ or } x \models^+ B,$$

$x \models^+ A{\rightarrow}B$ iff $\forall y \in M$ $[xRy$ and $y \models^+ A$ imply $y \models^+ B]$,

$x \models^+ {\sim}A$ iff $x \models^- A$,

$x \models^+ [m]A$ iff $x \models^m A$,

$x \models^+ [n]A$ iff $x \models^n A$,

$x \models^- A \wedge B$ iff $x \models^- A$ or $x \models^- B$,

$x \models^- A \vee B$ iff $x \models^- A$ and $x \models^- B$,

$x \models^- A{\rightarrow}B$ iff $x \models^+ A$ and $x \models^- B$,

$x \models^- {\sim}A$ iff $x \models^+ A$,

$x \models^- [m]A$ iff $x \models^n A$,

$x \models^- [n]A$ iff $x \models^m A$,

$x \models^m A \circ B$ iff $x \models^m A$ and $x \models^m B$, for $\circ \in \{\wedge, \vee, \rightarrow\}$,

$x \models^m \circ A$ iff $x \models^m A$, for $\circ \in \{\sim, [m], [n]\}$,

$x \models^n A \circ B$ iff $x \models^n A$ or $x \models^n B$, for $\circ \in \{\wedge, \vee, \rightarrow\}$,

$x \models^n \circ A$ iff $x \models^n A$, for $\circ \in \{\sim, [m], [n]\}$.

If $\mathcal{F} = \langle M, R \rangle$ is a Kripke frame, then $\langle M, R, \models^+, \models^-, \models^m, \models^n \rangle$ is a Kripke model *for* **N4mn** *based on* $\mathcal{F}$.

The heredity condition holds for $\models^+$, $\models^-$, $\models^m$, and $\models^n$, i.e., for any $\mathcal{L}$-formula A and any $x, y \in M$, if $x \models^* A$ and xRy, then $y \models^* A$, for $* \in \{+, -, m, n\}$.

As to a motivation of the semantical clauses for $[m]$ and $[n]$, we may note that a state supports the meaningfulness (nonsensicality) of a compound formula iff the state supports the meaningfulness (nonsensicality) of all (some) of its immediate proper subformulas; meaninglessness is 'infectious'. Thus, in particular, $x \models^m [n]A$ iff $x \models^m A$ iff $x \models^- [n]A$, and $x \models^m [n]A$ does not, in general, imply $x \models^+ [n]A$, although it does imply $x \models^- [n]A$. A state supports the meaningfulness of the statement that A is nonsensical iff the state supports the meaningfulness of A, and in this case the falsity of $[n]A$ is supported.

Definition 5 *An $\mathcal{L}$-formula A is said to be* true *in a Kripke model for* **N4mn** $\langle M, R, \models^+, \models^-, \models^m, \models^n \rangle$ *if* $x \models^+ A$ *for any* $x \in M$, *and to be* valid *on a Kripke frame* $\mathcal{F} = \langle M, R \rangle$ *if it is true for every Kripke model for* **N4mn** *based on* $\mathcal{F}$. *An $\mathcal{L}$-formula A is said to be* **N4mn**-*valid if A is valid on every Kripke frame. Let* $\Gamma \cup \{A\}$ *be a set of $\mathcal{L}$-formulas. Entailment is defined in terms of support-of-truth preservation at each state:* $\Gamma \models^+ A$ *if for all Kripke* *models for* **N4mn** $\langle M, R, \models^+, \models^-, \models^m, \models^n \rangle$ *and for all* $x \in M$, $x \models^+ A$ *if* $x \models^+ B$ *for all* $B \in \Gamma$. *We write* $A \models^+ B$ *for* $\{A\} \models^+ B$. *We define the logic* **N4mn** *model-theoretically as the pair* $\langle \mathcal{L}, \{\langle \Gamma, A \rangle \mid \Gamma \models^+ A\} \rangle$ *and* **N4** *is model-theoretically defined as* $\langle \mathcal{L}'', \{\langle \Gamma, A \rangle \mid \Gamma \models^+ A\} \rangle$.

Proposition (Wansing & Ayhan, 2023) *Each of the unary connectives* $\circ \in \{\sim, [m], [n]\}$ *is congruentiality-breaking in the sense that there are $\mathcal{L}$-formulas A and B such that* $A \models^+ B$ *and* $B \models^+ A$ *but not:* $\circ A \models^+ \circ B$ *and* $\circ B \models^+ \circ A$.

Definition 6 *Given the set Φ of propositional variables, we define three more sets of propositional variables, namely* $\Phi^- := \{p^- \mid p \in \Phi\}$, $\Phi^m := \{p^m \mid p \in \Phi\}$, *and* $\Phi^n := \{p^n \mid p \in \Phi\}$. *We inductively define a mapping f from* $\mathrm{Form}_{\mathcal{L}}(\Phi)$ *to the set of formulas of the language $\mathcal{L}'$ of* **IPL**$^+$ *defined over* $\Phi \cup \Phi^- \cup \Phi^m \cup \Phi^n$ *as follows:*

1. *for any* $p \in \Phi$, $f(p) = p$, $f(\sim p) = p^-$, $f([m]p) = p^m$, $f([n]p) = p^n$,

2. $f(A \circ B) = f(A) \circ f(B)$, *for* $\circ \in \{\rightarrow, \wedge, \vee\}$,

3. $f(\sim(A \wedge B)) = f(\sim A) \vee f(\sim B)$,

4. $f(\sim(A \vee B)) = f(\sim A) \wedge f(\sim B)$,

5. $f(\sim(A \rightarrow B)) = f(A) \wedge f(\sim B)$,

6. $f(\sim\sim A) = f(A)$,

7. $f(\sim[m]A) = f([n]A)$,

8. $f(\sim[n]A) = f([m]A)$,

9. $f([m](A \circ B)) = f([m]A) \wedge f([m]B)$, *for* $\circ \in \{\rightarrow, \wedge, \vee\}$,

10. $f([m] \circ A) = f([m]A)$, *for* $\circ \in \{\sim, [m], [n]\}$,

11. $f([n](A \circ B)) = f([n]A) \vee f([n]B)$, *for* $\circ \in \{\rightarrow, \wedge, \vee\}$,

12. $f([n] \circ A) = f([n]A)$, *for* $\circ \in \{\sim, [m], [n]\}$.

We write $f(\Gamma)$ to denote the result of replacing every formula A in Γ by $f(A)$; thus, $f(\varnothing) = \varnothing$.

Lemma 1 *Let f be the function defined in Definition 6. For any Kripke model for* **N4mn** $\langle M, R, \models^+, \models^-, \models^m, \models^n \rangle$, *we can define a Kripke model for* **Int**$^+$ $\langle M, R, \models \rangle$ *such that for any* $A \in \mathrm{Form}_{\mathcal{L}}(\Phi)$ *and any* $x \in M$,

1. $x \models^+ A$ *iff* $x \models f(A)$,

2. $x \models^- A$ iff $x \models f(\sim A)$,

3. $x \models^m A$ iff $x \models f([m]A)$,

4. $x \models^n A$ iff $x \models f([n]A)$.

Lemma 2 *Let f again be the function defined in Definition 6. Then for any Kripke model $\langle M, R, \models \rangle$ for* **IPL$^+$**, *we can construct a Kripke model $\langle M, R, \models^+, \models^-, \models^m, \models^n \rangle$ for* **N4mn** *such that for any $\mathcal{L}$-formula A and any $x \in M$,*

1. $x \models f(A)$ iff $x \models^+ A$,

2. $x \models f(\sim A)$ iff $x \models^- A$,

3. $x \models f([m]A)$ iff $x \models^m A$,

4. $x \models f([n]A)$ iff $x \models^n A$.

Theorem 1 (Semantical embedding) (Wansing and Ayhan (2023)) *Let f be the mapping from Definition 6. For any set of $\mathcal{L}$-formulas $\Gamma \cup A$, $\Gamma \models^+ A$ in* **N4mn** *iff $f(\Gamma) \models f(A)$ in* **IPL$^+$**.

4 The logic $\mathbb{Inf}$ of the information order of the lattice 16_{inf}

In this section, I will introduce another logic in a language that contains next to negation, $\sim$, the one-place sentential operators $[m]$ and $[n]$, expressing meaningfulness, respectively nonsensicality. This system, $\mathbb{Inf}$, is a many-valued logic that is arrived at by (i) translating the support of truth, support of falsity, support of meaningfulness, and support of nonsensicality conditions for $\sim$, $[m]$, and $[n]$ in **N4mn** into truth tables and (ii) interpreting conjunction and disjunction as the lattice meet and lattice join of a certain lattice of generalized semantical values. As a result, in $\mathbb{Inf}$ the connectives $\sim$, $[m]$, $[n]$, conjunction, and disjunction interact differently from how they interact in **N4mn**. I will keep the notation for the meaningfulness and nonsensicality connectives but use the notation for fusion and fission known from, for example, the logic of logical bilattices for conjunction, respectively disjunction. In contrast to the language of **N4mn**, the language of $\mathbb{Inf}$ contains no (primitive) conditional.

4.1 The semantics of $\mathbb{Inf}$

The propositional language $\mathcal{L}_{inf}$ contains the unary connectives $\sim$, $[m]$, $[n]$ and the binary connectives $\otimes$ (fusion) and $\oplus$ (fission) over a denumerable set

Heinrich Wansing

Φ of propositional variables. The set $Form(\mathcal{L}_{inf})$ of $\mathcal{L}_{inf}$-formulas over Φ is defined in the standard way. For the semantics, our starting point is the set $4 = \{\mathbf{t}, \mathbf{f}, \mathbf{m}, \mathbf{n}\}$. We read $\mathbf{t}$ as "true", $\mathbf{f}$ as "false", $\mathbf{m}$ as "meaningful", and $\mathbf{n}$ as "nonsensical". We generalize these basic values to obtain 'told-values' and consider the sixteen elements of 16, the powerset $\mathcal{P}(4)$ of 4:

1. $\varnothing$ (told neither true nor false nor meaningful nor nonsensical)
2. $\{\mathbf{t}\}$ (told only true)
3. $\{\mathbf{f}\}$ (told only false)
4. $\{\mathbf{m}\}$ (told only meaningful)
5. $\{\mathbf{n}\}$ (told only nonsensical)
6. $\{\mathbf{t}, \mathbf{f}\}$ (told both true and false)
7. $\{\mathbf{t}, \mathbf{m}\}$ (told both true and meaningful)
8. $\{\mathbf{t}, \mathbf{n}\}$ (told both true and nonsensical)
9. $\{\mathbf{f}, \mathbf{m}\}$ (told both false and meaningful)
10. $\{\mathbf{f}, \mathbf{n}\}$ (told both false and nonsensical)
11. $\{\mathbf{m}, \mathbf{n}\}$ (told both meaningful and nonsensical)
12. $\{\mathbf{t}, \mathbf{f}, \mathbf{m}\}$ (told true, false, and meaningful)
13. $\{\mathbf{t}, \mathbf{f}, \mathbf{n}\}$ (told true, false, and nonsensical)
14. $\{\mathbf{t}, \mathbf{m}, \mathbf{n}\}$ (told true, meaningful, and nonsensical)
15. $\{\mathbf{f}, \mathbf{m}, \mathbf{n}\}$ (told false, meaningful, and nonsensical)
16. $\{\mathbf{t}, \mathbf{f}, \mathbf{m}, \mathbf{n}\}$ (told true, false, meaningful, and nonsensical).

If we order 16 by set-inclusion, we obtain the distributive complete lattice $16_{inf} = (16, \subseteq)$, which is depicted as a Hasse-diagram in Figure 1.

A valuation in 16 is a function $v^{16}\colon \Phi \longrightarrow 16$. Valuation functions v^{16} in 16 are extended to functions from $Form(\mathcal{L}_{inf})$ to 16 as follows:

$$v^{16}(A \otimes B) = v^{16}(A) \cap v^{16}(B)$$
$$v^{16}(A \oplus B) = v^{16}(A) \cup v^{16}(B)$$

$\mathbf{t} \in v^{16}(\sim A)$ iff $\mathbf{f} \in v^{16}(A)$
$\mathbf{f} \in v^{16}(\sim A)$ iff $\mathbf{t} \in v^{16}(A)$
$\mathbf{m} \in v^{16}(\sim A)$ iff $\mathbf{m} \in v^{16}(A)$
$\mathbf{n} \in v^{16}(\sim A)$ iff $\mathbf{n} \in v^{16}(A)$

$\mathbf{t} \in v^{16}([m]A)$ iff $\mathbf{m} \in v^{16}(A)$
$\mathbf{f} \in v^{16}([m]A)$ iff $\mathbf{n} \in v^{16}(A)$
$\mathbf{m} \in v^{16}([m]A)$ iff $\mathbf{m} \in v^{16}(A)$
$\mathbf{n} \in v^{16}([m]A)$ iff $\mathbf{n} \in v^{16}(A)$

$\mathbf{t} \in v^{16}([n]A)$ iff $\mathbf{n} \in v^{16}(A)$
$\mathbf{f} \in v^{16}([n]A)$ iff $\mathbf{m} \in v^{16}(A)$
$\mathbf{m} \in v^{16}([n]A)$ iff $\mathbf{m} \in v^{16}(A)$
$\mathbf{n} \in v^{16}([n]A)$ iff $\mathbf{n} \in v^{16}(A)$.

174

Definition 7 (Informational entailment)   *The entailment relation $\models^{16}_i \subseteq$* $(Form(\mathcal{L}_{inf}) \times Form(\mathcal{L}_{inf}))$ *is defined by setting*

$$A \models^{16}_i B \text{ iff for every valuation } v^{16} \text{ in } 16, v^{16}(A) \subseteq v^{16}(B).$$

Definition 8   *The logic* $\mathbb{Inf}$ *is presented syntactically as the relation $\vdash_i \subseteq$* $(Form(\mathcal{L}_{inf}) \times Form(\mathcal{L}_{inf}))$ *defined by the following axiomatic statements and rules, where* $\circ \in \{\sim, [m], [n]\}$, $\sharp \in \{\otimes, \oplus\}$, *and* $A \dashv\vdash_i B$ *is a shorthand for* $A \vdash_i B$ *and* $B \vdash_i A$:

Axioms	$[m] \circ A \dashv\vdash_i [m]A$	([m]∘ *reduction*)
	$[n] \circ A \dashv\vdash_i [n]A$	([n]∘ *reduction*)
	$[m](A\sharp B) \dashv\vdash_i [m]A\sharp[m]B$	([m]♯ *distribution*)
	$[n](A\sharp B) \dashv\vdash_i [n]A\sharp[n]B$	([n]♯ *distribution*)
	$A \otimes B \vdash_i A, \ A \otimes B \vdash_i B$	(⊗-*elim*)
	$A \vdash_i A \oplus B, \ B \vdash_i A \oplus B$	(⊕-*intro*)
	$A \otimes (B \oplus C) \vdash_i (A \otimes B) \oplus (A \otimes C)$	(*distribution*)
	$\sim\sim A \dashv\vdash_i A$	(*double negation*)
	$\sim[m]A \dashv\vdash_i [n]A$	(*negated* [m])
	$\sim[n]A \dashv\vdash_i [m]A$	(*negated* [n])
	$\sim(A \otimes B) \dashv\vdash_i \sim A \otimes \sim B$	(*negated fusion*)
	$\sim(A \oplus B) \dashv\vdash_i \sim A \oplus \sim B$	(*negated fission*)

Rules	$A \vdash_i B$ and $A \vdash_i C$ together imply $A \vdash_i B \otimes C$	(⊗-*intro*)
	$A \vdash_i C$ and $B \vdash_i C$ together imply $A \oplus B \vdash_i C$	(⊕-*elim*)
	$A \vdash_i B$ and $B \vdash_i C$ together imply $A \vdash C$	(*transitivity*)

Note that

1. $$A \vdash_i A \qquad \text{(reflexivity)}$$

   is derivable by (double negation) and (transitivity),

2. from the axioms and rules for $\otimes$ and $\oplus$ it is clear that there is sense in which $A \otimes B$, respectively $A \oplus B$, is a conjunction, respectively disjunction, and

3. $$A \vdash_i B \text{ implies } \sim B \vdash_i \sim A \qquad \text{(contraposition)}$$

   is not validity preserving.

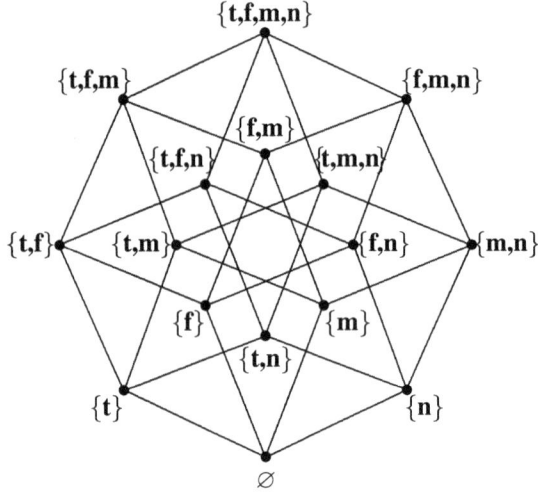

Figure 1: The lattice $\mathbb{16}_{inf}$

The failure of (contraposition) is as it should be if truth and falsity are two independent semantical dimensions in their own right and (contraposition) is not explicitly imposed by definition on a negation connective, as it is usually the case in the study of logics resulting from bi- or tri- or other multilattices.

To prove completeness we will construct a suitable canonical model, see (Shramko & Wansing, 2005; 2011). Let $\alpha \subseteq Form(\mathcal{L}_{inf})$. Then α is a *theory* if

- if $A \in \alpha$ and $A \vdash_i B$, then $B \in \alpha$,
- if $A \in \alpha$ and $B \in \alpha$, then $A \otimes B \in \alpha$.

A theory α is said to be *prime* iff $A \oplus B \in \alpha$ implies that $A \in \alpha$ or $B \in \alpha$. The following fact about prime theories is very well known, a proof is given, for example, in (Dunn, 2000, p. 13):

Lemma 3 *For any A and $B \in Form(\mathcal{L}_{inf})$, if $A \nvdash_i B$, then there exists a prime theory α such that $A \in \alpha$ and $B \notin \alpha$.*

For any prime theory α we define the canonical valuation $v_{\mathcal{T}} \colon \Phi \longrightarrow \mathbb{16}$ as follows:

$$\mathbf{t} \in v_{\mathcal{T}}(p) \text{ iff } p \in \alpha; \qquad\qquad \mathbf{f} \in v_{\mathcal{T}}(p) \text{ iff } \sim p \in \alpha;$$
$$\mathbf{m} \in v_{\mathcal{T}}(p) \text{ iff } [m]p \in \alpha; \qquad\qquad \mathbf{n} \in v_{\mathcal{T}}(p) \text{ iff } [n]p \in \alpha.$$

Remarks on Semantic Information and Logic

Canonical valuations can be extended to arbitrary $Form(\mathcal{L}_{inf})$-formulas.

Lemma 4 *Let α be a prime theory and let v_T be defined as above. Then for any $A \in Form(\mathcal{L}_{inf})$:*

$\mathbf{t} \in v_T(A)$ *iff* $A \in \alpha$; $\mathbf{f} \in v_T(A)$ *iff* $\sim\!A \in \alpha$;

$\mathbf{m} \in v_T(A)$ *iff* $[m]A \in \alpha$; $\mathbf{n} \in v_T(A)$ *iff* $[n]A \in \alpha$.

Proof. The proof is by induction on the construction of formulas $A \in Form(\mathcal{L}_{inf})$. For propositional variables the claim holds by definition.

If A has the form $\sim\!B$, then we have $\mathbf{t} \in v_T(\sim\!B)$ iff $\mathbf{f} \in v_T(B)$ iff, by the induction hypothesis, $\sim\!B \in \alpha$; $\mathbf{m} \in v_T(\sim\!B)$ iff $\mathbf{m} \in v_T(B)$ iff, by the induction hypothesis, $[m]B \in \alpha$ iff $[m]\sim\!B \in \alpha$ by ($[m]\sim$ reduction); $\mathbf{f} \in v_T(\sim\!B)$ iff $\mathbf{t} \in v_T(B)$ iff, by the induction hypothesis, $B \in \alpha$ iff, by (double negation), $\sim\!\sim\!B \in \alpha$; $\mathbf{n} \in v_T(\sim\!B)$ iff $\mathbf{n} \in v_T(B)$ iff, by the induction hypothesis, $[n]B \in \alpha$ iff, by ($[n]\sim$ reduction), $[n]\sim\!B \in \alpha$.

If A has the form $[m]B$, then $\mathbf{t} \in v_T([m]B)$ iff $\mathbf{m} \in v_T(B)$ iff, by the induction hypothesis, $[m]B \in \alpha$; $\mathbf{m} \in v_T([m]B)$ iff $\mathbf{m} \in v_T(B)$ iff, by the induction hypothesis, $[m]B \in \alpha$ iff $[m][m]B \in \alpha$ by ($[m][m]$ reduction); $\mathbf{f} \in v_T([m]B)$ iff $\mathbf{n} \in v_T(B)$ iff, by the induction hypothesis, $[n]B \in \alpha$ iff, by (negated $[m]$), $\sim\![m]B \in \alpha$; $\mathbf{n} \in v_T([m]B)$ iff $\mathbf{n} \in v_T(B)$ iff, by the induction hypothesis, $[n]B \in \alpha$ iff, by ($[n][m]$ reduction), $[n][m]B \in \alpha$.

If A has the form $[n]B$, then $\mathbf{t} \in v_T([n]B)$ iff $\mathbf{n} \in v_T(B)$ iff, by the induction hypothesis, $[n]B \in \alpha$; $\mathbf{m} \in v_T([n]B)$ iff $\mathbf{m} \in v_T(B)$ iff, by the induction hypothesis, $[m]B \in \alpha$ iff $[m][n]B \in \alpha$ by ($[m][n]$ reduction); $\mathbf{f} \in v_T([n]B)$ iff $\mathbf{m} \in v_T(B)$ iff, by the induction hypothesis, $[m]B \in \alpha$ iff, by (negated $[n]$), $\sim\![n]B \in \alpha$; $\mathbf{n} \in v_T([n]B)$ iff $\mathbf{n} \in v_T(B)$ iff, by the induction hypothesis, $[n]B \in \alpha$ iff, by ($[n][n]$ reduction), $[n][n]B \in \alpha$.

If A has the form $B \otimes C$, we have $\mathbf{t} \in v_T(B \otimes C)$ iff $\mathbf{t} \in v_T(B) \cap v_T(C)$ iff ($\mathbf{t} \in (v_T(B)$ and $\mathbf{t} \in v_T(C))$ iff, by the induction hypothesis, ($B \in \alpha$ and $C \in \alpha$) iff, by the definition of theories and ($\otimes$-elim), $B \otimes C \in \alpha$; $\mathbf{m} \in v_T(B \otimes C)$ iff ($\mathbf{m} \in (v_T(B)$ and $\mathbf{m} \in v_T(C))$ iff, by the induction hypothesis, ($[m]B \in \alpha$ and $[m]C \in \alpha$) iff, by the definition of theories and ($\otimes$-elim), $[m]B \otimes [m]C \in \alpha$ iff $[m](B \otimes C) \in \alpha$ by ($[m]\otimes$ distribution). Next, $\mathbf{f} \in v_T(B \otimes C)$ iff $\mathbf{f} \in v_T(B) \cap v_T(C)$ iff ($\mathbf{f} \in (v_T(B)$ and $\mathbf{f} \in v_T(C))$ iff, by the induction hypothesis, ($\sim\!B \in \alpha$ and $\sim\!C \in \alpha$) iff, by the definition of theories, ($\otimes$-elim), and (negated fusion), $\sim\!(B \otimes C) \in \alpha$; $\mathbf{n} \in v_T(B \otimes C)$ iff ($\mathbf{n} \in (v_T(B)$ and $\mathbf{n} \in v_T(C))$ iff, by the induction hypothesis, ($[n]B \in \alpha$ and $[n]C \in \alpha$) iff, by the definition of theories and ($\otimes$-elim), $[n]B \otimes [n]C \in \alpha$ iff $[n](B \otimes C) \in \alpha$ by ($[n]\otimes$ distribution).

If A has the form $B \oplus C$, the reasoning is analogous to that of the previous case and makes use of (negated fission), ($[m]\oplus$ distribution), ($[n]\oplus$ distribution), ($\oplus$-intro), and the definition of prime theories. □

The proof of the characterization theorem for $\vdash_i$ follows a standard pattern.

Theorem 2 *For any $A, B \in Form(\mathcal{L}_{inf})$: $A \models_i^{16} B$ iff then $A \vdash_i B$.*

Proof. Right-to-left (soundness): It is easy to show that the axioms are valid and that the rules preserve validity. For the (negated $[m]$) axioms, for example, we have

$$\begin{array}{ll}
\mathbf{t} \in v^{16}(\sim[m]A) & \text{iff} \\
\mathbf{f} \in v^{16}([m]A) & \text{iff} \\
\mathbf{n} \in v^{16}(A) & \text{iff} \\
\mathbf{t} \in v^{16}([n]A) &
\end{array} \qquad \begin{array}{ll}
\mathbf{f} \in v^{16}(\sim[m]A) & \text{iff} \\
\mathbf{t} \in v^{16}([m]A) & \text{iff} \\
\mathbf{m} \in v^{16}(A) & \text{iff} \\
\mathbf{f} \in v^{16}([n]A) &
\end{array}$$

$$\begin{array}{ll}
\mathbf{m} \in v^{16}(\sim[m]A) & \text{iff} \\
\mathbf{m} \in v^{16}([m]A) & \text{iff} \\
\mathbf{m} \in v^{16}(A) & \text{iff} \\
\mathbf{m} \in v^{16}([n]A) &
\end{array} \qquad \begin{array}{ll}
\mathbf{n} \in v^{16}(\sim[m]A) & \text{iff} \\
\mathbf{n} \in v^{16}([m]A) & \text{iff} \\
\mathbf{n} \in v^{16}(A) & \text{iff} \\
\mathbf{n} \in v^{16}([n]A). &
\end{array}$$

Left-to-right (completeness): Let $A \models_i^{16} B$ and assume $A \nvdash B$. By Lemma 3, there exists a prime theory α such that $A \in \alpha$ and $B \notin \alpha$. Then, by Lemma 4, $\mathbf{t} \in v_\mathcal{T}(A)$ but $\mathbf{t} \notin v_\mathcal{T}(B)$, and thus $A \nvDash_i^{16} B$. (Likewise, we can consider $\mathbf{f}$ instead of $\mathbf{t}$. Let $A \in \alpha$ and $B \notin \alpha$. Then, by (double negation), this is the case iff $\sim\sim A \in \alpha$ and $\sim\sim B \notin \alpha$. By Lemma 4, $\mathbf{f} \in v_\mathcal{T}(\sim A)$ but $\mathbf{f} \notin v_\mathcal{T}(\sim B)$, and thus $A \nvDash_i^{16} B$.) □

4.2 From the lattice 16_{inf} to the pentalattice 65536_5

It is, of course, possible to define further partial orderings on 16 in addition to the subset relation, but I will define additional orderings on $65536 = \mathcal{P}(16)$ instead of 16. The reason for this is similar to the reason for considering the trilattice $SIXTEEN_3$ instead of the bilattice $FOUR_2$. Both $FOUR_2$ and $SIXTEEN_3$ give rise to a semantics for the basic propositional paraconsistent relevance logic known as Belnap-Dunn logic, Dunn-Belnap logic, or first-degree entailment logic, **FDE**, see (Anderson & Belnap, 1975, § 15.2), or, for a survey and additional references, (Omori & Wansing, 2017). The language of **FDE** contains the connectives $\sim$ (negation), $\wedge$ (conjunction), and $\vee$ (disjunction), and **FDE** can be defined as the logic of what is usually said to be the *truth order*, $\leq_t$, of $FOUR_2$. The bilattice $FOUR_2$ is defined on a set of four semantical values, **T**, **F**, **N**, and **B**, which have the following intuitive reading, stated already in Section 1:

T (*told true but not false*)
F (*told false but not true*)
N (*told neither true nor false*)
B (*told both true and false*).

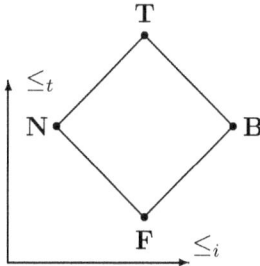

Figure 2: The bilattice $FOUR_2$.

In Figure 2, the bilattice $FOUR_2$ with its two partial orders is depicted as a Hasse diagram. The values **T**, **F**, **N**, and **B** can be represented as the elements of the powerset $\mathcal{P}(\{T, F\}) = 4$ of the set of classical truth values $2 = \{T, F\}$: $\mathbf{N} = \varnothing$, $\mathbf{T} = \{T\}$, $\mathbf{F} = \{F\}$, $\mathbf{B} = \{T, F\}$, see (Dunn, 1976; 2000). With this representation, the information order $\leq_i$ on 4 is the subset relation.

First-degree entailment logic is semantically determined by interpreting conjunction and disjunction as the lattice meet, respectively lattice join of $\leq_t$. Negation is interpreted by a unary operation, $-$, that inverts the truth order, leaves the information order untouched and satisfies $x = --x$. Alternatively, the semantics can be given by the matrix $\langle 4, \{\mathbf{T}, \mathbf{B}\}, \{f_c : c \in \{\sim, \wedge, \vee\}\}\rangle$, where the functions f_c are defined by the following truth tables:

$f_\sim$	
T	**F**
B	**B**
N	**N**
F	**T**

$f_\wedge$	**T**	**B**	**N**	**F**
T	**T**	**B**	**N**	**F**
B	**B**	**B**	**F**	**F**
N	**N**	**F**	**N**	**F**
F	**F**	**F**	**F**	**F**

$f_\vee$	**T**	**B**	**N**	**F**
T	**T**	**T**	**T**	**T**
B	**T**	**B**	**T**	**B**
N	**T**	**T**	**N**	**N**
F	**T**	**B**	**N**	**F**

The set $\mathcal{D} = \{\mathbf{T}, \mathbf{B}\}$ is the set of designated values. A valuation function v mapping propositional variables into $\mathbf{4}$ is extended to a valuation of arbitrary formulas by requiring that $v(c(A_1, \ldots, A_m)) = f_c(v(A_1), \ldots, v(A_m))$, and the semantic consequence relation $\models_{\mathbf{FDE}}$ between single formulas A and B is defined as follows:

$A \models_{\mathbf{FDE}} B$ iff for every valuation function v, $v(A) \in \mathcal{D}$ implies $v(B) \in \mathcal{D}$

or, equivalently, by setting

$$A \models_{\mathbf{FDE}} B \text{ iff for every valuation function } v, v(A) \leq_t v(B).$$

The relation $\leq_i$ on $\mathbf{4}$ can quite convincingly be seen as an information ordering, the idea being that the more elements a semantical value contains the more informative is the assignment of that value to a propositional variable. In (Shramko & Wansing, 2005; 2011) it is argued that it is much less convincing to regard $\leq_t$ as a truth ordering. The reason for viewing $\{T\}$ as 'more true' than $\{T, F\}$ and regarding $\varnothing$ as 'more true' than $\{F\}$ is the absence of the classical value F from $\{T\}$, respectively $\varnothing$. The relation $\leq_t$ is thus not defined only with respect to the presence of T in or the absence of T from elements of $\mathbf{4}$. If one moves from $\mathbf{4}$ to $\mathcal{P}(\mathbf{4}) = \mathbf{16}$, however, it is possible not only to define a pure truth ordering in terms of the presence of T in or the absence of T from elements of elements of $\mathbf{16}$ but also a pure falsity ordering in terms of the presence of F in or the absence of F from elements of elements of $\mathbf{16}$.

In (Shramko & Wansing, 2005; 2011) in addition to the subset relation as an information order on $\mathbf{16}$, a truth and a falsity ordering are defined. In a first step, for every x in $\mathbf{16}$ the sets x^t, x^{-t}, x^f, and x^{-f} are defined by the following equations:

$$x^t := \{y \in x \mid T \in y\}; \qquad x^{-t} := \{y \in x \mid T \notin y\};$$
$$x^f := \{y \in x \mid F \in y\}; \qquad x^{-f} := \{y \in x \mid F \notin y\}.$$

Definition 9 *For every x, y in* $\mathbf{16}$:

- $x \leq_i y$ *iff* $x \subseteq y$;
- $x \leq_t y$ *iff* $x^t \subseteq y^t$ *and* $y^{-t} \subseteq x^{-t}$;
- $x \leq_f y$ *iff* $x^f \subseteq y^f$ *and* $y^{-f} \subseteq x^{-f}$.

Remarks on Semantic Information and Logic

Following that strategy, for every x in 65536 we define the sets x^t, x^{-t}, x^f, x^{-f}, x^m, x^{-m}, x^n, and x^{-n} as follows:

$$x^t := \{y \in x \mid \mathbf{t} \in y\};\qquad x^{-t} := \{y \in x \mid \mathbf{t} \notin y\};$$
$$x^f := \{y \in x \mid \mathbf{f} \in y\};\qquad x^{-f} := \{y \in x \mid \mathbf{f} \notin y\};$$
$$x^m := \{y \in x \mid \mathbf{m} \in y\};\qquad x^{-m} := \{y \in x \mid \mathbf{m} \notin y\};$$
$$x^n := \{y \in x \mid \mathbf{n} \in y\};\qquad x^{-n} := \{y \in x \mid \mathbf{n} \notin y\}.$$

Definition 10 *For every x, y in 65536:*

- $x \leq_i y$ iff $x \subseteq y$;
- $x \leq_\circ y$ iff $x^\circ \subseteq y^\circ$ and $y^{-\circ} \subseteq x^{-\circ}$, for $\circ \in \{t, f, m, n\}$.

Lattice meet and lattice join operations for all five partial orderings exist, and we will denote them as $X \sqcap_\circ Y$, respectively $X \sqcup_\circ Y$ for $\circ \in \{t, f, m, n\}$. With this definition, we obtain the pentalattice

$$65536_5 = (65536, \subseteq, \leq_t, \leq_f, \leq_m, \leq_n).$$

The pentalattice 65536_5 gives rise to the propositional language $\mathcal{L}(65536_5)$ based on a denumerable set of propositional variables Φ defined in Backus-Naur form as follows:

variables Φ: $\quad p \in \Phi$
formulas: $\quad A \in Form_{(\mathcal{L}_{65536_5})}(\Phi)$
$A ::= \quad p \mid (A \otimes A) \mid (A \oplus A) \mid (A \wedge_t A) \mid (A \vee_t A) \mid (A \wedge_f A) \mid (A \vee_f A) \mid$
$\quad (A \wedge_m A) \mid (A \vee_m A) \mid (A \wedge_n A) \mid (A \vee_n A) \mid \sim A \mid [m]A \mid [n]A.$

Let $\mathbf{t}^* := \mathbf{f}$, $\mathbf{f}^* := \mathbf{t}$, $\mathbf{m}^* := \mathbf{m}$, $\mathbf{n}^* := \mathbf{n}$, $X^* := \{x^* \mid x \in X\}$ for $X \in 16$, and $X^* := \{X^* \mid X \in X\}$ for $X \in 65536$. Let $\mathbf{t}^m := \mathbf{m}$, $\mathbf{f}^m := \mathbf{n}$, $\mathbf{m}^m := \mathbf{m}$, $\mathbf{n}^m := \mathbf{n}$, $X^m := \{x^m \mid x \in X\}$ for $X \in 16$, and $X^m := \{X^m \mid X \in X\}$ for $X \in 65536$. Let $\mathbf{t}^n := \mathbf{n}$, $\mathbf{f}^n := \mathbf{m}$, $\mathbf{m}^n := \mathbf{m}$, $\mathbf{n}^n := \mathbf{n}$, $X^n := \{x^n \mid x \in X\}$ for $X \in 16$, and $X^n := \{X^n \mid X \in X\}$ for $X \in 65536$. A valuation in 65536 is a function $v^\odot \colon \Phi \longrightarrow 65536$. Valuation functions $v^\odot$ in 65536 are extended to functions from $Form_{(\mathcal{L}_{65536_5})}(\Phi)$ to 65536 as follows, where $\circ \in \{t, f, m, n\}$:

$$
\begin{aligned}
v^\odot(A \otimes B) &= v^\odot(A) \cap v^\odot(B); & v^\odot(\sim A) &= (v^\odot(A))^*; \\
v^\odot(A \oplus B) &= v^\odot(A) \cup v^\odot(B); & v^\odot([m]A) &= (v^\odot(A))^m; \\
v^\odot(A \wedge_\circ B) &= v^\odot(A) \sqcap_\circ v^\odot(B); & v^\odot([n]A) &= (v^\odot(A))^n. \\
v^\odot(A \vee_\circ B) &= v^\odot(A) \sqcup_\circ v^\odot(B); &
\end{aligned}
$$

Heinrich Wansing

Definition 11 *The relation* $\models_i^{65536} \subseteq (Form(\mathcal{L}_{inf}) \times Form(\mathcal{L}_{inf}))$ *is defined by setting*

$A \models_i^{65536} B$ *iff for every valuation* $v^\odot$ *in* 65536, $v^\odot(A) \subseteq v^\odot(B)$.

Conjecture *For any* $A, B \in Form(\mathcal{L}_{inf})$: $A \models_i^{65536} B$ *iff* $A \vdash_i B$.

I expect no particular obstacle to verifying the conjecture. An anonymous referee wondered whether it could be proved by some embedding of the lattice 16 into the lattice 65536_5.

5 Negation inconsistency

It is well known that a certain simple modification of the support of falsity condition for implications in **N4** leads to a non-trivial negation inconsistent connexive logic, namely the system **C**, see (Wansing, 2005), (Omori & Wansing, 2020), (Niki & Wansing, 2023). The same modification brings us from **N4mn** to the connexive logic **Cmn**. The notion of a Kripke model for **Cmn** and the notion of **Cmn**-validity are defined in analogy to the case of **N4mn**, except that the following falsification clause for implications $A \to B$ is used: $x \models^- A{\to}B$ iff $\forall y \in M$ [xRy and $y \models^+ A$ imply $y \models^- B$].

Assuming this falsification clause, the logic **Cmn** is model-theoretically defined as the pair $\langle \mathcal{L}, \{\langle \Gamma, A \rangle \mid \Gamma \models^+ A\} \rangle$.

Theorem 3 (Semantical embedding) *Let* f' *be the mapping from* $Form_\mathcal{L}(\Phi)$ *to the set of formulas of the language* $\mathcal{L}'$ *defined over* $\Phi \cup \Phi^- \cup \Phi^m \cup \Phi^n$ *that is defined exactly like the function* f *from Definition 6, except that*

$$f'(\sim(A \to B)) = f'(A) \to f'(\sim B).$$

Then, for any set of $\mathcal{L}$-*formulas* $\Gamma \cup A$, $\Gamma \models^+ A$ *in* **Cmn** *iff* $f'(\Gamma) \models f'(A)$ *in* **IPL**$^+$.

The following schematic formulas, for instance, are **Cmn**-valid:

$$(A \to (\sim A \to A)) \text{ and } \sim(A \to (\sim A \to A)).$$

The language $\mathcal{L}_{inf}$ of Inf does not contain a genuine conditional, by which I mean an implication connective, $\to$, that satisfies the deduction theorem, i.e., validates implication introduction, and modus ponens. The addition of any such conditional to Inf presented as a relation $\vdash_i$ between sets of $\mathcal{L}_{inf}$-formulas and single $\mathcal{L}_{inf}$-formulas will result in a nontrivial negation inconsistent logic. Given that $\varnothing \vdash_i A \to A$ is provable, we get the following derivations:

1. $\varnothing \vdash_i A \to A$
2. $A \to A \vdash_i \sim(A \to A) \oplus (A \to A)$ $\oplus$-introduction
3. $\varnothing \vdash_i \sim(A \to A) \oplus (A \to A)$ 1., 2., transitivity

1. $\varnothing \vdash_i A \to A$
2. $A \to A \vdash_i \sim\sim(A \to A)$ double negation
3. $\varnothing \vdash_i \sim\sim(A \to A)$ 1., 2., transitivity
4. $\sim\sim(A \to A) \vdash_i \sim\sim(A \to A) \oplus \sim(A \to A)$ $\oplus$-introduction
5. $\varnothing \vdash_i \sim\sim(A \to A) \oplus \sim(A \to A)$ 3., 4., transitivity
6. $\sim\sim(A \to A) \oplus \sim(A \to A) \vdash_i \sim(\sim(A \to A) \oplus (A \to A))$ negated fission
7. $\varnothing \vdash_i \sim(\sim(A \to A) \oplus (A \to A))$ 5., 6., transitivity

6   Open problems and further directions

In addition to deciding the above conjecture, open questions and directions for future research abound. First, in addition to $\mathcal{L}_{inf}$ we can define the following fragments $\mathcal{L}_\circ$ of $\mathcal{L}(65536_5)$ based on a denumerable set of propositional variables Φ, for $\circ \in \{t, f, m, n\}$:

variables Φ: $p \in \Phi$
formulas: $A \in Form_{\mathcal{L}_\circ}(\Phi)$
$A ::= \ p \mid (A \wedge_\circ A) \mid (A \vee_\circ A) \mid \sim A \mid [m]A \mid [n]A.$

Naturally, for each of the languages $\mathcal{L}(65536_5)$ and $\mathcal{L}_\circ$, and not only for $\mathcal{L}_{inf}$, we can define informational entailment:

$A \models_i^{65536} B$ iff for every valuation $v^\odot$ in 65536, $v^\odot(A) \subseteq v^\odot(B)$.

Next, for each of the languages $\mathcal{L}(65536_5)$ and $\mathcal{L}_\circ$, we can define the following entailment relations:

Definition 12   *The entailment relation* $\models_\circ^{65536} \subseteq (Form_{(\mathcal{L}_{65536_5})}(\Phi) \times Form_{(\mathcal{L}_{65536_5})}(\Phi))$ *is defined by setting*

$A \models_\circ^{65536} B$ *iff for every valuation* $v^\odot$ *in* 65536, $v^\odot(A) \leq_\circ v^\odot(B)$.

The relation $\models_{\mathcal{L}_\circ}^{65536} \subseteq (Form(\mathcal{L}_\circ)(\Phi) \times Form(\mathcal{L}_\circ)(\Phi))$ *is defined by setting*

$A \models_{\mathcal{L}_\circ}^{65536} B$ *iff for every valuation* $v^\odot$ *in* 65536, $v^\odot(A) \leq_\circ v^\odot(B)$.

Moreover, we can think of combinations of those relations, e.g., by considering intersections. Since all theses languages lack a primitive conditional, it makes a lot of sense to expand the languages with a genuine implication connective. An obvious task then is to define proof systems for the various semantically introduced logics with nice proof-theoretic properties, especially sequent calculi that allow for proof analysis.

References

Anderson, A., & Belnap, N. (1975). *Entailment. Volume 1*. Princeton, NJ: Princeton University Press.

Belnap, N. D. (1976). How a computer should think. In G. Ryle (Ed.), *Contemporary Aspects of Philosophy* (pp. 30–56). Stocksfield: Oriel Press.

Belnap, N. D. (1977). A useful four-valued logic. In M. Dunn & G. Epstein (Eds.), *Modern Uses of Multiple-valued Logic* (pp. 5–37). Dordrecht: Reidel.

Dunn, J. M. (1976). Intuitive semantics for first-degree entailment and 'coupled trees'. *Philosophical Studies, 29*, 149–168.

Dunn, J. M. (2000). Partiality and it dual. *Studia Logica, 66*, 5–40.

Floridi, L. (2010). *Information. A Very Short Introduction*. Oxford: Oxford University Press.

Grzegorczyk, A. (1964). A philosophically plausible formal interpretation of intuitionistic logic. *Indagationes Mathematicae, 26*, 596–601.

Kripke, S. (1965). Semantical analysis of intuitionistic logic I. In J. Crossley & M. Dummett (Eds.), *Formal Systems and Recursive Functions* (pp. 92–130). Amsterdam: North-Holland Publishing Company.

Mares, E. (2008). Information, negation, and paraconsistency. In F. Berto, E. Mares, K. Tanaka, & F. Paoli (Eds.), *Paraconsistency: Logic and Applications* (pp. 43–55). Dordrecht: Springer.

Mares, E. (2009). General information in relevant logic. *Synthese, 167*, 343–362.

Mares, E. (2010). The nature of information: A relevant approach. *Synthese, 175 (supplement 1)*, 111–132.

Niki, S., & Wansing, H. (2023). On the provable contradictions of the connexive logics **C** and **C3**. *Journal of Philosophical Logic, 52(5)*, 355–1383.

Odintsov, S. P. (2008). *Constructive Negations and Paraconsistency.* Dodrecht: Springer.

Omori, H., & Wansing, H. (2017). 40 years of FDE: An Introductory Overview. *Studia Logica, 105,* 1021–1049.

Omori, H., & Wansing, H. (2020). An extension of connexive logic **C**. In N. Olivetti, R. Verbrugge, S. Negri, & G. Sandu (Eds.), *Advances in Modal Logic. Vol. 13* (pp. 503–522). London: College Publications.

Punčochář, V. (2016). Algebras of information states. *Journal of Logic and Computation, 27,* 1643–1675.

Punčochář, V. (2019). Substructural inquisitive logics. *Review of Symbolic Logic, 12,* 296–330.

Punčochář, V., & Sedlár, I. (2021). Epistemic extensions of substructural inquisitive logics. *Journal of Logic and Computation, 31,* 1820–1844.

Shramko, Y., & Wansing, H. (2005). Some useful 16-valued logics: How a computer network should think. *Journal of Philosophical Logic, 34,* 121–153.

Shramko, Y., & Wansing, H. (2011). *Truth and Falsehood. An Inquiry into Generalized Logical Values.* Dordrecht: Springer.

Shramko, Y., & Wansing, H. (2021). The nature of entailment: an informational approach. *Synthese, 198,* 5241–5261.

Urquhart, A. (1972). Semantics for relevant logics. *Journal of Symbolic Logic, 37,* 159–169.

Wansing, H. (1993a). Informational interpretation of substructural propositional logics. *Journal of Logic, Language and Information, 2,* 285–308.

Wansing, H. (1993b). *The Logic of Information Structures.* Berlin: Springer.

Wansing, H. (2005). Connexive Modal Logic. In R. Schmidt, I. Pratt-Hartmann, M. Reynolds, & H. Wansing (Eds.), *Advances in Modal Logic. Volume 5* (pp. 367–383). London: King's College Publications. See also ⟨http://www.aiml.net/volumes/volume5/⟩.

Wansing, H. (2022). One heresy and one orthodoxy: On dialetheism, dimathematism, and the non-normativity of logic. *Erkenntnis,* published online 09 March 2022.

Wansing, H., & Ayhan, S. (2023). Logical multilateralism. *Journal of Philosophical Logic,* published online, 25 September.

Wansing, H., & Odintsov, S. (2016). On the methodology of paraconsistent logic. In H. Andreas & P. Verdée (Ed.), *Logical Studies of Paraconsistent Reasoning in Science and Mathematics* (pp. 175–204). Dordrecht: Springer.

Weiss, Y. (2022). Semantics for pure theories of connexive implication. *The Review of Symbolic Logic*, *15*, 591–606.

Heinrich Wansing
Department of Philosophy I, Ruhr University Bochum
Germany
E-mail: `Heinrich.Wansing@rub.de`

www.ingramcontent.com/pod-product-compliance
Lightning Source LLC
Chambersburg PA
CBHW060517090426

42735CB00011B/2263